SIGNATURE

OF

CONTROVERSY

SIGNATURE OF CONTROVERSY

RESPONSES TO CRITICS OF
SIGNATURE IN THE CELL

DAVID KLINGHOFFER, EDITOR

DISCOVERY INSTITUTE PRESS   SEATTLE   2010

Description

This book gathers previously published essays and blog posts responding to criticism of *Signature in the Cell: DNA and the Evidence for Intelligent Design*, by Stephen C. Meyer. Contributors include Stephen Meyer, David Berlinski, Richard Sternberg, and Jay Richards. Edited by David Klinghoffer.

Copyright Notice

Copyright © 2010 by Discovery Institute. All Rights Reserved.

Publisher's Note

This book is part of a series published by the Center for Science & Culture at Discovery Institute in Seattle. Previous books include *The Deniable Darwin & Other Essays*, by David Berlinski; *Darwin's Conservatives: The Misguided Quest*, by John West; and *Traipsing into Evolution: Intelligent Design and the Kitzmiller vs. Dover Decision*, by David DeWolf et al.

Library Cataloging Data

Signature of Controversy: Responses to Critics of Signature in the Cell
Edited by David Klinghoffer.

143 pages, 6 x 9 x 0.35 inches & 0.5 lb, 229 x 152 x 2.29 cm. & 0.29 kg

BISAC Subject: SCI027000   SCIENCE / Life Sciences / Evolution
BISAC Subject: SCI080000   SCIENCE / Essays
BISAC Subject: SCI034000   SCIENCE / History

ISBN-13: 978-0-9790141-8-5   ISBN-10: 0-9790141-8-2 (paperback)

Publisher Information

Discovery Institute Press, 208 Columbia Street, Seattle, WA 98101
Internet: http://www.discoveryinstitutepress.com/
Published in the United States of America on acid-free paper.
First Edition, First Printing. August 2010.

Contents

Introduction	7
Signature in the Cell: The Executive Summary	11
I. On Not Reading Stephen Meyer's *Signature in the Cell*	15
1. On Not Reading *Signature in the Cell*: A Response to Francisco Ayala	17
2. When a Book Review Is Not a "Book Review"	27
3. Falk's Rejoinder to Meyer's Response to Ayala's "Essay" on Meyer's Book	31
4. Lying for Darwin	37
5. Responding to Stephen Fletcher in the *Times Literary Supplement*	39
6. Responding Again to Stephen Fletcher in the *Times Literary Supplement*	43
7. Responding to Stephen Fletcher in the *Times Literary Supplement*	47
8. Why Are Darwinists Scared to Read *Signature in the Cell*?	51
9. Every Bit Digital: DNA's Programming Really Bugs Some ID Critics	55
10. Ayala: "For the Record, I Read *Signature in the Cell*"	57
II. On Reading Stephen Meyer's *Signature in the Cell*	59
11. Responding to Darrel Falk's Review of *Signature in the Cell*	61
12. Asking Darrel Falk to Pick a Number, Any Number	69
13. Ayala and Falk Miss the Signs in the Genome	71
14. Discovering Signs in the Genome by Thinking Outside the BioLogos Box	77
15. Beginning to Decipher the SINE Signal	83
16. Intelligent Design, Frontloading and Theistic Evolution	89
17. Getting ID Right: Further Thoughts on the Beliefnet Review	93
III. Attack of the Pygmies	97
18. Signs of Desperation? Early Responses to *Signature in the Cell*	99
19. Get Smart: Can Unintelligent Causes Produce Biological Information?	105
20. Weather Forecasting and Complex Specified Information	111
21. Gotcha! Stephen Meyer's Spelling & Other Weighty Criticisms	117
22. Matheson's Intron Fairy Tale	125
23. Let's Do the Math Again	129
24. Darwinian Tree-Huggers: You Gotta Love Their Devotion	131
25. Is Intelligent Design Bad Theology?	135
About the Authors	141

Introduction

David Klinghoffer

PUBLISHED IN 2009, STEPHEN MEYER'S *SIGNATURE IN THE CELL: DNA and the Evidence for Intelligent Design* has already been recognized as establishing one of the strongest pillars underlying the argument for intelligent design. This massive and massively original work needs to be read and studied by every thoughtful person who cares about what is surely the ultimate question facing every human being: Where did life come from? Charles Darwin himself did not seek to resolve that mystery, but his modern followers believe the problem is well in hand, along Darwinian lines, offering a variety of purely materialist explanations for the origin of the biological information coded in DNA. Meyer masterfully sweeps aside all such guesses and assumptions and demonstrates that science points to an origin of life emanating from somewhere or someone outside nature.

To call Meyer's book fascinating and important is an understatement. No less interesting in its way, however, was the critical response and it is with that the book you are reading now is concerned. For the fact is that despite its being written about in print and online by numerous friends and foes of intelligent-design theory, few—if any—of the critics really grappled with the substance of Meyer's argument. This is remarkable and telling.

In the pages that follow, which include links to the critics' own writings, defenders of Stephen Meyer's book analyze the hostile response. The chapters here all appeared previously, most on the Discovery Institute's group blog site, Evolution News and Views (ENV), on the BioLogos and Biologic Institute sites, or in the journal Salvo. The book is organized along the following lines. In Part I, Meyer and his defenders go to work on the horde of *Signature*-bashers who not only did not read the book but in most instances did not even take the trouble to inform themselves about its contents. These latter include even so eminent a biologist as Francisco Ayala of the University of California, Irvine—of whom, more in a moment. For the convenience of the busy Darwinist on the go, with no time to read a book he plans to attack and virtually no time even to read about it, we precede Part I with a helpful Executive Summary of *Signature in the Cell*. It is hoped this will avert future embarrassments like the review that Professor Ayala contributed to the BioLogos website.

In Part II, Meyer and other friends of ID reply to critics who actually took the time to read *Signature in the Cell* before attacking it. This turned out to be a relative rarity, for reasons that are worth pondering. While Parts I and II deal with *Signature*'s more serious critics, or anyway those with reputations for seriousness, Part III concentrates on the crowd of pygmies who populate the furious, often obscene world of Darwinist and other blogs.

Admittedly, in editing this volume, it was not always obvious to me which critics belong under which heading. For example, Jerry Coyne is a University of Chicago biologist who lately seems to spend most of his time blogging. Yet he clearly belongs among the ranks of the more distinguished writers who bashed Meyer's book without reading it or reading about it. On the other hand, such an individual as blogger Jeffrey Shallit, mathematician at the University of Waterloo in Ontario, Canada—not to be confused with the University of Wallamaloo of Monty Python fame—may object to being classed as a pygmy. Oh well. Sorry.

READERS OF THIS book may wonder why the essays and blog posts collected here include many responses to critics who attacked *Signature in the Cell* without having read it. Wouldn't it be more illuminating to engage solely with those who are at least adequately familiar with what Stephen Meyer wrote? The truth is, it was necessary both to write and to collect and publish these defenses because some of the most prominent attacks were precisely from scientists who did not read the book but felt entitled to comment anyway. This fact is important because it illustrates the difficulty faced by the intelligent-design community in seeking to get a fair hearing. Thus, a point worth repeating, the aforementioned Francisco Ayala critiqued *Signature in the Cell* at length *despite having virtually no idea what is in it*. Let that sink in.

It's funny, or maybe just sad. A couple of years ago I wrote an article for *Townhall* magazine about the suppression of intelligent-design advocates in university and other academic settings. At the time I was writing it, I sent an email to several prominent theistic evolutionists and other Darwin defenders, including Dr. Ayala. I asked:

> Critics of ID argue that one failing of ID theory, among others, is that it hasn't been backed up by research. If you were to imagine a university-employed scientist who wanted to do such research, would he be completely free to do so? Or, as ID advocates say, would he more likely be dissuaded by pressure from peers or supervisors?

Ayala replied:

> He would be free to do so. I cannot imagine any serious scientist or academic administrator trying to dissuade anybody else from carrying out any well-designed research project (or, in fact, almost any research project). Our academic freedom to pursue any research we wish is something precious that we value as much as any other academic value.

Well, that is just rich. After the experiences of Sternberg, Gonzalez, Crocker, Marks, Minnich, Dembski, Coppedge—chronicled on ENV and elsewhere, along with other suppressed scientists yet to be named and still others too worried about reprisals to let themselves to be identified—we know Ayala's statement to be utterly false. When it comes to publicly doubting Darwin, serious scientists would be justified in feeling intimidated. In part, the fear of speaking out is maintained by the realization that if you raise your voice, your view will not merely be criticized. It will be distorted so as to prejudice public and professional opinion against you.

What we have in the Ayala affair, a genuine scandal, is a telling illustration of how that works.

SIGNATURE IN THE CELL:
THE EXECUTIVE SUMMARY

> Why a summary of *Signature*? As noted in several of the essays that follow, one motif in critical responses to Dr. Meyer has been the fact that many hostile reviewers had not read the book and were not even aware what was in it. Well, that's understandable. Just as the busy, stressed-out executive in the business world needs his information boiled down to essentials for him, perhaps so too does the frantically harried defender of scientific materialism. With future critics in mind, we thought it would be helpful to provide the following capsule summary. —EDITOR

UNTIL NOW, THE MODERN INTELLIGENT DESIGN MOVEMENT HAS focused principally on the explanatory inadequacies of Darwin's mechanism to account for evolutionary novelty. In *Signature in the Cell*, Stephen C. Meyer sets aside questions pertaining to biological evolution to resolve an even more fundamental mystery, one that Darwin made no attempt to address: How did life begin?

Meyer investigates contemporary theories of the origin of life. He reports on the reasons why the hypothesis depending upon chance has been almost universally abandoned, while also examining modes of explanation based on natural law, and those based on a combination of chance and necessity. Concluding that such theories cannot adequately explain a fundamental property of living systems—their ability to harbor and utilize functional information content—Meyer argues that no contemporary materialist theory provides a plausible picture of the origin of life. Using a methodology of scientific reasoning pioneered by Charles Darwin himself, Meyer fleshes out a positive scientific case for intelligent design—one based upon what he describes as our "uniform and repeated experience" of the cause-and-effect structure of the world. In the entire universe, there is only one cause known to be capable of producing large volumes of complex specified information. And that is rational deliberation, or purposive design.

Meyer also evaluates the often-heard complaint that ID does not make any testable predictions. He presents a dozen such predictions made by ID, with a detailed commentary on the functional character of so-called "junk DNA" complete with a comprehensive citation of literature for functions recently assigned to various non-coding regions of the genome including its role in gene regulation. The latter function Meyer likens to the operating system in a computer.

Meyer draws to a close by considering the question of "Why It Matters." He includes a discussion of the theological issues at stake in the debate over origins, as well as an overview of the larger worldview implications of detectable design in nature. He differentiates between the evidential basis—the premises—of a theory and its philosophical implications and consequences.

In contrast to the common stereotype of science where scientific enquiry is presumed to be worldview-neutral, many scientific theories—in particular origins theories—have wider philosophical implications. Such consequences should not be considered adequate grounds to dismiss a theory as "unscientific." Take the Big Bang as an example, a theory in cosmology that points to the temporal nature of the cosmos. Many scientists initially rejected the theory because of the challenge it posed to the idea, favored by materialists, of an eternal, unchanging universe. Meyer also points to Darwinism as a scientific theory with potential implications far beyond the realm of objective scientific enquiry.

The epilogue provides a discussion of the potential of the ID paradigm to drive scientific research in new and fruitful directions, proposing research questions that might be tackled from the perspective of ID. Meyer describes what he terms "extragenomic or ontogenetic information," arguing that much of the information responsible for the development of organismal form does not reside at the DNA level. DNA does not determine how individual proteins assemble themselves into larger systems of proteins, nor does it determine how cell types, tissue types, and organs are arranged into bodies. Given that, it would seem DNA sequences can mutate indefinitely and still never produce a fundamentally new body plan.

Meyer suggests that the activity of a rational, purposive agent be considered a plausible explanation for the origin and development of life, taking into account what we already know about the unique ability of such an agent to design information-rich parts and to organize those parts into functional information-rich systems and hierarchies.

I

ON NOT READING STEPHEN MEYER'S *SIGNATURE IN THE CELL*

1. On Not Reading *Signature in the Cell*: A Response to Francisco Ayala

Stephen C. Meyer

Dr. Francisco Ayala, professor of biological sciences, ecology and evolutionary biology, as well as of logic and the philosophy of science, at the University of California, Irvine, reviewed *Signature in the Cell* for the BioLogos Foundation's website.[1] Below is Dr. Meyer's response. —Editor

No doubt it happens all the time. There must be many book reviews written by reviewers who have scarcely cracked the pages of the books they purport to review. But those who decide to write such blind reviews typically make at least some effort to acquire information about the book in question so they can describe its content accurately—if for no other reason than to avoid embarrassing themselves. Unfortunately, in his review of my book *Signature in the Cell* (titled ironically, "On *Reading* the Cell's Signature"), eminent evolutionary biologist Francisco Ayala does not appear to have even made a search for the crib notes online. Indeed, from reading his review on the BioLogos website it appears that he did little more than glance at the title page and table of contents—if that. As a result, his review misrepresents the thesis and topic of the book and even misstates its title.

The title of my book is not *Signature of the Cell* as Ayala repeatedly refers to it, but *Signature in the Cell*.

The thesis of the book is not that "chance, by itself, cannot account for the genetic information found in the genomes of organisms" as he claims, but instead that intelligent design *can* explain, and does provide the best explanation for (among many contenders, not just chance) the origin of the information necessary to produce the first living cell.

1. http://biologos.org/blog/on-reading-the-cells-signature/

Further, the topic that the book addresses is not the origin of the genomes of organisms or the *human* genome as the balance of Professor Ayala's critique seems to imply, but instead the origin of the *first* life and the mystery surrounding the origin of the information necessary to produce it.

Ayala begins his review by attempting to trivialize the argument of *Signature in the Cell*. But he does so by misrepresenting its thesis. According to Ayala, "The keystone argument of *Signature of the Cell* [sic] is that chance, by itself, cannot account for the genetic information found in the genomes of organisms." He notes—as I do in the book—that all evolutionary biologists already accept that conclusion. He asks: "Why, then, spend chapter after chapter and hundreds of pages of elegant prose to argue the point?" But, of course, the book does not spend hundreds of pages arguing that point. In fact, it spends only 55 pages out of 613 pages explaining why origin-of-life researchers have—since the 1960s—almost universally come to reject the chance hypothesis. It does so, not because the central purpose of the book is to refute the chance hypothesis per se, but for several other reasons intrinsic to the actual thesis of the book.

Signature in the Cell makes a case for the design hypothesis as the *best* explanation for the origin of the biological information necessary to produce the first living organism. In so doing, it deliberately employs a standard method of historical scientific reasoning, one that Darwin himself affirmed and partly pioneered in the *Origin of Species*. The method, variously described as the method of multiple competing hypotheses or the method of inferring to the best explanation, necessarily requires an examination of the main competing hypotheses that scientists have proposed to explain a given event in the remote past. Following Darwin and his scientific mentor Lyell, historical scientists have understood that *best* explanations typically cite causes that are known from present experience to be capable, indeed uniquely capable, of producing the effect in question.

IN THE PROCESS of using the method of multiple competing hypotheses to develop my case for intelligent design in *Signature in the Cell*, I do examine the chance hypothesis for the origin of life, because it is one of the many competing hypotheses that have been proposed to explain the origin of the first life and the origin of biological information. Naturally, since chance was one of the first hypotheses proposed to explain the origin of life in the wake of the discovery of the information-bearing properties of DNA, I critique it first. Nevertheless, I go on to examine many more recent models for the origin of biological information including those that rely on physical-chemical necessity (such as current self-organizational models), and those that rely on the interplay between chance and necessity (such as the popular RNA world scenario). My discussion of these models takes over ninety pages and four chapters. Did Ayala just miss these chapters?

I should add that my critique of the chance hypothesis provides a foundation for assessing some of these more recent chemical evolutionary theories—theories that Ayala would presumably recognize as contenders among contemporary evolutionary biologists and which rely on chance in combination with other processes. For example, in the currently popular RNA world scenario, self-replicating RNA catalysts are posited to have first arisen as the result of random interactions between the chemical building blocks or subunits of RNA. According to advocates of this view, once such self-replicating RNA molecules had come into existence, then natural selection would have become a factor in the subsequent process of molecular evolution necessary to produce the first cell. In *Signature in the Cell*, however, I show that the amount of sequence-specific information necessary to produce even a supposedly simple self-replicating RNA molecule far exceeds what can be reasonably assumed to have arisen by chance alone. Indeed, my analysis of the probabilities of producing various information-rich bio-molecules is not only relevant to showing that "chance, by itself, cannot account for" the origin of genetic information, but also to showing why theories that invoke chance in combination with pre-biotic natural selection likewise fail.

In any case, *Signature in the Cell* does not just make a case *against* materialistic theories for the origin of the information necessary to produce the first life, it also makes a positive case *for* intelligent design by showing that the activity of conscious and rational agents is the only known cause by which large amounts of new functional information arise, at least when starting from purely physical and chemical antecedents.

The closest that Ayala comes in his review to recognizing the central affirmative argument in the book is his rather clumsy attempt to refute the idea of intelligent design by insisting that the existence of "nonsensical" or junk sequences in the human genome demonstrates that it did not arise by intelligent design. As he claims explicitly, "according to Meyer, ID provides a more satisfactory explanation of the human genome than evolution does."

Again, I have to wonder whether Professor Ayala even cracked the pages of the book. My book is not about the origin of the *human* genome, nor about human evolution nor even biological evolution generally. It's about chemical evolution, the origin of the first life and the genetic information necessary to produce it. In fact, I explicitly acknowledge in the epilogue that someone could in principle accept my argument for the intelligent design of the first life and also accept the standard neo-Darwinian account of how subsequent forms of life evolved. I don't hold this "front-end loaded" view of design, but my book makes no attempt to refute it or standard accounts of biological evolution. For this reason, it's hard to see how Ayala's attempt to defend biological evolution and refute the particular hypothesis that intelligent design played a discernible role in the origin of the human genome in any way challenges the argument of *Signature in the Cell*.

Even so, it is worth noting that the argument that Ayala makes against intelligent design of the human genome based upon on the presence of "nonsensical" or so-called junk DNA is predicated upon two factually flawed and out-of-date premises. Ayala suggests that no designer worthy of the modifier "intelligent" would have allowed the human genome to be liberally sprinkled with a preponderance of nonsense DNA sequences and that the presence and apparently random distribution of such sequences is more adequately explained as a by-product of the trial and error process of undirected mutation and selection. According to Ayala, the distribution of a particular sequence (the *Alu* sequence), which he asserts contains genetic nonsense, suggests a sloppy, unintelligent editor, not an intelligent designer. As he argues:

> It is as if the editor of *Signature of the Cell* would have inserted between every two pages of Meyer's book, forty additional pages, each

containing the same three hundred letters. Likely, Meyer would not think of his editor as being "intelligent." Would a function ever be found for these one million nearly identical *Alu* sequences? It seems most unlikely.

Thus, in essence, Ayala claims that (1) a preponderance of nonsense DNA sequences and (2) the random distribution of these sequences shows that the human genome could not have been intelligently designed. But both of the factual claims upon which Ayala bases this argument are wrong.

First, neither the human genome nor the genomes of other organisms are predominantly populated with junk DNA. As I document in *Signature in the Cell*, the non-protein-coding regions of the genomes (of various organisms) that were long thought to be "junk" or "nonsense" are now known to perform numerous mission-critical functions. Non-protein-coding DNA is neither nonsense nor junk. On page 407 of *Signature in the Cell*, I enumerate ten separate functions that non-protein-coding regions of the genome are now known to play. (References to peer-reviewed scientific publications documenting my claims are provided there). Overall, the non-coding regions of the genome function much like the operating system in a computer in that they direct and regulate the timing and expression of the other protein-coding genetic modules.

Further, the *Alu* sequences that Ayala specifically cites as prime examples of widely and randomly distributed nonsense sequences in the human genome are *not* non-functional or "nonsense." Short Interspersed Nuclear Element (SINE) sequences, of which *Alu* is one member, perform numerous formatting and regulatory functions in the genomes of all organisms in which they have been found. It is simply factually incorrect for Ayala to claim otherwise.

IN GENERAL, SINEs (and thus *Alus*) allow genetic information to be retrieved in multiple different ways from the same DNA data files depending on the specific needs of different cell types or tissues (in different species-specific contexts). In particular, *Alu* sequences perform many taxon-specific lower-level genomic formatting functions such as: (1) providing alternative start sites for promoter modules in gene expression—somewhat like sectoring on a hard drive (Faulkner et al., 2009; Faulkner and Carninci, 2009); (2) suppressing or "silencing" RNA transcription (Trujillo et al., 2006); (3) dynamically partitioning one gene file from another on the chromosome (Lunyak et al., 2007); (4) providing DNA nodes for signal transduction pathways or binding sites for hormone receptors (Jacobsen et al., 2009; Laperriere et al., 2004); (5) encoding RNAs that modulate transcription (Allen et al., 2004; Espinoza et al., 2004; Walters et al., 2009); and (6) encoding or regulating microRNAs (Gu et al., 2009; Lehnert et al., 2009).

In addition to these lower-level genomic formatting functions, SINEs (including *Alus*) also perform species-specific higher-level genomic formatting functions such as: (1) modulating the chromatin of classes of GC-rich housekeeping and signal transduction genes (Grover et al., 2003, 2004; Oei et al., 2004; see also Eller et al., 2007); (2) "bar coding" particular segments for chromatin looping between promoter and enhancer elements (Ford and Thanos, 2010); (3) augmenting recombination in sequences where *Alus* occur (Witherspoon et al., 2009); and (4) assisting in the formation of three-dimensional chromosome territories or "compartments" in the nucleus (Kaplan et al., 1993; see also Pai and Engelke, 2010).

Moreover, *Alu* sequences also specify many species-specific RNA codes. In particular, they provide: (1) signals for alternative RNA splicing (i.e., they generate multiple messenger RNAs from the same type of precursor transcript) (Gal-Mark et al., 2008; Lei and Vorechovsky, 2005; Lev-Maor et al., 2008) and (2) alternative open-reading frames (exons) (Lev-Maor et al., 2007; Lin et al., 2008; Schwartz et al., 2009). *Alu* sequences also (3) specify the retention of select RNAs in the nucleus to silence expression (Chen et al., 2008; Walters et al., 2009); (4) regulate the RNA polymerase II machinery during transcription (Mariner et al., 2008; Yakovchuk et al., 2009; Walters et al., 2009); and (5) provide sites for Adenine-to-Inosine RNA editing, a function that is essential for both human development and species-specific brain development (Walters et al., 2009).

CONTRARY TO AYALA'S claim, *Alu* sequences (and other mammalian SINEs) are not distributed randomly but instead manifest a similar "bar code" distribution pattern along their chromosomes (Chen and Manuelidis, 1989; Gibbs et al., 2004; Korenberg and Rykowski, 1988). Rather like the distribution of the backslashes, semi-colons and spaces involved in the formatting of software code, the "bar code" distribution of *Alu* sequences (and other SINEs) reflects a clear functional logic, not sloppy editing or random mutational insertions. For example, *Alu* sequences are preferentially located in and around protein-coding genes as befits their role in regulating gene expression (Tsirigos and Rigoutsos, 2009). They occur mainly in promoter regions—the start sites for RNA production—and in introns, the segments that break up the protein-coding stretches. Outside of these areas, the numbers of *Alu* sequences sharply decline. Further, we now know that *Alu* sequences are directed to (or spliced into) certain preferential hotspots in the genome by the protein complexes or the "integrative machinery" of the cell's information processing system (Levy et al., 2010). This directed distribution of *Alu* sequences enhances the semantic and syntactical organization of human DNA. It appears to have little to do with the occurrence of random insertional mutations, contrary to the implication of Ayala's "sloppy editor" illustration and argument.

Critics repeatedly claim that the theory of intelligent design is based on religion, not science. But in his response to my book, it is Ayala who relies on a theological argument and who repeatedly misrepresents the scientific literature in a vain attempt to support it. The human genome manifests nonsense sequences and sloppy editing ill-befitting of a deity or any truly *intelligent* designer, he argues. He also sees other aspects of the natural world that he thinks are inconsistent with the existence of a Deity. I'll leave it to theologians to grapple with Ayala's arguments about whether backaches in old age and other forms of generalized human suffering make the existence of God logically untenable. But on the specific scientific question of the organization of the human genome, I think the evidence is clear. It is Ayala who has been sloppy, and not only in his assessment of the human genome, but also, I must add, in his critique of my book.

Bibliography

Allen, T. A., S. Von Kaenel, J. A. Goodrich and J. F. Kugel, "The SINE-encoded mouse B2 RNA represses mRNA transcription in response to heat shock," *Nature Structural and Molecular Biology* 11:9 (2004), pp. 816–821.

Chen, L. L., J. N. DeCerbo and G. G. Carmichael, "*Alu* element-mediated gene silencing," *EMBO Journal* 27:12 (2008), pp. 1694–1705.

Chen, T. L. and L. Manuelidis, "SINEs and LINEs cluster in distinct DNA fragments of Giemsa band size," *Chromosoma* 98:5 (1989), pp. 309–316.

Eller, C. D., M. Regelson, B. Merriman, S. Nelson, S. Horvath and Y. Marahrens, "Repetitive sequence environment distinguishes housekeeping genes," *Gene* 390:1-2 (2007), pp. 153–165.

Espinoza, C. A., T. A. Allen, A. R. Hieb, J. F. Kugel and J. A. Goodrich, "B2 RNA binds directly to RNA polymerase II to repress transcript synthesis," *Nature Structural and Molecular Biology* 11:9 (2004), pp. 822–829.

Faulkner, G. J. and P. Carninci, "Altruistic functions for selfish DNA," *Cell Cycle* 8:18 (2009), pp. 2895–2900.

Faulkner, G. J., Y. Kimura, C. O. Daub, S. Wani, C. Plessy, K. M. Irvine, K. Schroder, N. Cloonan, A. L. Steptoe, T. Lassmann, K. Waki, N. Hornig, T. Arakawa, H. Takahashi, J. Kawai, A. R. Forrest, H. Suzuki, Y. Hayashizaki, D. A. Hume, V. Orlando, S. M. Grimmond and P. Carninci, "The regulated retrotransposon transcriptome of mammalian cells," *Nature Genetics* 41:5 (2009), pp. 563–571.

Ford, E. and D. Thanos (in press), "The transcriptional code of human IFN-beta gene expression," *Biochimica et Biophysica Acta* (2010).

Gal-Mark, N., S. Schwartz and G. Ast, "Alternative splicing of *Alu* exons—two arms are better than one," *Nucleic Acids Research* 36:6 (2008), pp. 2012–2023.

Gibbs, R. A., G. M. Weinstock, M. L. Metzker, D. M. Muzny et al., "Genome sequence of the Brown Norway rat yields insights into mammalian evolution," *Nature* 428:6982 (2004), pp. 493–521.

Grover, D., P. P. Majumder, B. C. Rao, S. K. Brahmachari and M. Mukerji, "Nonrandom distribution of *alu* elements in genes of various functional categories: insight from analysis of human chromosomes 21 and 22," *Molecular Biology and Evolution* 20:9 (2003), pp. 1420–1424.

Grover, D., M. Mukerji, P. Bhatnagar, K. Kannan and S. K. Brahmachari, "*Alu* repeat analysis in the complete human genome: trends and variations with respect to genomic composition," *Bioinformatics* 20:6 (2004), pp. 813–817.

Gu, T. J., X. Yi, X. W. Zhao, Y. Zhao and J. Q. Yin, "*Alu*-directed transcriptional regulation of some novel miRNAs," *BMC Genomics* 10:563 (2009).

Jacobsen, B. M., P. Jambal, S. A. Schittone and K. B. Horwitz, "*Alu* repeats in promoters are position-dependent co-response elements (coRE) that enhance or repress transcription by dimeric and monomeric progesterone receptors," *Molecular Endocrinology* 23:7 (2009), pp. 989–1000.

Kaplan, F. S., J. Murray, J. E. Sylvester, I. L. Gonzalez, J. P. O'Connor, J. L. Doering, M. Muenke, B. S. Emanuel and M. A. Zasloff, "The topographic organization of repetitive DNA in the human nucleolus," *Genomics* 15:1 (1993), pp. 123–132.

Korenberg, J. R. and M. C Rykowski, "Human genome organization: *Alu*, lines, and the molecular structure of metaphase chromosome bands," *Cell* 53:3 (1988), pp. 391–400.

Laperriere, D. T. T. Wang, J. H. White and S. Mader, "Widespread *Alu* repeat-driven expansion of consensus DR2 retinoic acid response elements during primate evolution," *BMC Genomics* 8:23 (2004).

Lehnert, S., P. Van Loo, P. J. Thilakarathne, P. Marynen, G. Verbeke and F. C. Schuit, "Evidence for co-evolution between human microRNAs and *Alu*-repeats," *PLoS One* 4:2 (2009), e4456.

Lei, H. and I. Vorechovsky, "Identification of splicing silencers and enhancers in sense *Alus*: a role for pseudoacceptors in splice site repression," *Molecular Cell Biology* 25:16 (2005), pp. 6912–6920.

Lev-Maor, G., O. Ram, E. Kim, N. Sela, A. Goren, E. Y. Levanon and G. Ast, "Intronic *Alus* influence alternative splicing," *PLoS Genetics* 4:9 (2008), e1000204.

Lev-Maor, G., R. Sorek, E. Y. Levanon, N. Paz, E. Eisenberg and G. Ast, "RNA-editing-mediated exon evolution," *Genome Biology* 8:2 (2007), R29.

Levy, A., S. Schwartz and G. Ast (in press), "Large-scale discovery of insertion hotspots and preferential integration sites of human transposed elements," *Nucleic Acids Research* (2010).

Lin, L., S. Shen, A. Tye, J. J. Cai, P. Jiang, B. L. Davidson and Y. Xing, "Diverse splicing patterns of exonized *Alu* elements in human tissues," *PLoS Genetics* 4:10 (2008), e1000225.

Lunyak. V. V., G. G. Prefontaine, E. Núñez, T. Cramer, B. G. Ju, K.A. Ohgi, K. Hutt, R. Roy, A. García-Díaz, X. Zhu, Y. Yung, L. Montoliu, C. K. Glass and M. G. Rosenfeld, "Developmentally regulated activation of a SINE B2 repeat as a domain boundary in organogenesis," *Science* 317:5835 (2007), pp. 248–251.

Mariner, P. D., R. D. Walters, C. A. Espinoza, L. F. Drullinger, S. D. Wagner, J. F. Kugel and J. A. Goodrich, "Human *Alu* RNA is a modular transacting repressor of mRNA transcription during heat shock," *Molecular Cell* 29:4 (2008), pp. 499–509.

Oei, S. L., V. S. Babich, V. I. Kazakov, N. M. Usmanova, A. V. Kropotov and N. V. Tomilin, "Clusters of regulatory signals for RNA polymerase II transcription associated with *Alu* family repeats and CpG islands in human promoters," *Genomics* 83:5 (2007), pp. 873–882.

Pai, D. A. and D. R. Engelke, "Spatial organization of genes as a component of regulated expression," *Chromosoma* 119:1 (2010), pp. 13–25.

Schwartz, S., N. Gal-Mark, N. Kfir, R. Oren, E. Kim and G. Ast, "*Alu* exonization events reveal features required for precise recognition of exons by the splicing machinery, "*PLoS Computational Biology* 5:3 (2009), e1000300.

Trujillo, M. A., M. Sakagashira and N. L. Eberhardt, "The human growth hormone gene contains a silencer embedded within an *Alu* repeat in the 3'-flanking region," *Molecular Endocrinology* 20:10 (2006), pp. 2559–2575.

Tsirigos, A. and I. Rigoutsos, "*Alu* and b1 repeats have been selectively retained in the upstream and intronic regions of genes of specific functional classes," *PLoS Computational Biology* 5:12 (2009), e1000610.

Walters, R. D., J. F. Kugel and J. A. Goodrich, "Inv*Alu*able junk: the cellular impact and function of *Alu* and B2 RNAs," *IUBMB Life* 61:8 (2009), pp. 831–837.

Witherspoon, D. J., W. S. Watkins, Y. Zhang, J. Xing, W. L. Tolpinrud, D. J. Hedges, M. A. Batzer and L. B. Jorde, "*Alu* repeats increase local recombination rates," *BMC Genomics* 10:530 (2009).

Yakovchuk, P., J. A. Goodrich and J. F. Kugel, "B2 RNA and *Alu* RNA repress transcription by disrupting contacts between RNA polymerase II and promoter DNA within assembled complexes," *Proceedings National Academy of Science USA* 106:14 (2009), pp. 5569-5574.

2. When a Book Review Is Not a "Book Review"

David Klinghoffer

As a former book review editor (at *National Review*), I take a professional interest in book reviews and all the things that can go right or wrong with them. I confess, though, I've never seen anything quite like the treatment of Stephen Meyer's book, *Signature in the Cell: DNA and the Evidence for Intelligent Design*, on BioLogos, the curious website specializing in Christian apologetics for Darwin. The site published what was clearly, unambiguously written to look like a review by biologist Francisco Ayala[2] that, as Steve Meyer pointed out[3] already, actually gave every evidence that Ayala had not read the book. (My colleague Dr. Meyer thinks Ayala *did* read the Table of Contents, but on this I must disagree.)

On what did Ayala base his views about *Signature*? This is a bit of a mystery. BioLogos president Dr. Darrel Falk is unstinting with fulsome praise for Ayala ("one of Biology's living legends"). Falk claims[4] he actually asked Ayala to respond to Falk's review of *Signature*.[5] Falk purports that in publishing Ayala's review, he mistakenly failed to introduce it with the disclaimer that Ayala was reviewing Falk's review, not Meyer's book per se. Yeah, sure. Falk's review did not provide Ayala with his absurd misrepresentation of Meyer's argument. Instead Ayala gives every impression of having derived that from his own assessment of the book itself. As Ayala claims,

> The keystone argument of *Signature of the Cell* [sic] is that chance, by itself, cannot account for the genetic information found in the genomes of organisms. I agree. And so does every evolutionary scien-

2. http://biologos.org/blog/on-reading-the-cells-signature/
3. http://biologos.org/blog/on-not-reading-the-signature-stephen-c-meyers-response-to-francisco-ayala-part-1/
4. http://biologos.org/blog/on-not-reading-the-signature-stephen-c-meyers-response-to-francisco-ayala-part-1/
5. http://biologos.org/blog/on-not-reading-the-signature-stephen-c-meyers-response-to-francisco-ayala-part-1/

tist, I presume. Why, then, spend chapter after chapter and hundreds of pages of elegant prose to argue the point?

Yet that is certainly *not* the keystone argument of *Signature*, and Meyer in fact spends only 55 pages (out of 613) on it. But that is not really the point here.

What's notable is that Falk in his own review,[6] whatever its other faults or merits, never claimed that *Signature* is all about proving that "chance, by itself, cannot, account for the genetic information found in genomes." Falk doesn't mention the word "chance." So where did Ayala get his mistaken notion? All one can say is, not from the book, which he patently didn't read, and not from Falk. Indeed, Ayala in his essay does not mention Falk or Falk's review. Clearly, Ayala wanted readers to *think* he was reviewing *Signature in the Cell*—or *Signature* of *the Cell* as he repeatedly calls it. Thus, for example, he commends Meyer for his "elegant prose." The idea that Ayala was merely acting in good faith on Falk's assignment of responding to Falk's review is hardly believable.

Okay, so far we have a reviewer reviewing a book he did not read and a book review editor (Falk's apparent role here) claiming disingenuously that it was all an innocent mix-up, that the review by the "living legend" was never intended as a review and was merely presented as one by mistake, even though it clearly reads like a review or critique or a critical evaluation—call it what you will.

On top of this, there is Falk's introduction to Meyer's response to Ayala. Here he essentially ambushes Meyer by agreeing to publish his reply to Ayala and then introduces the reply, in italics above it and at some length, in a blatant and again disingenuous attempt to undercut its credibility. Thus Falk claims that Meyer originally agreed to limit himself in his response to "Ayala's philosophical and theological arguments." In Falk's presentation, Meyer then stabbed *him* in the back by going ahead and writing about the science after all. In reality, in his full response, Meyer writes about philosophy (multiple competing hypotheses), theology (Ayala's claims about junk DNA), and science. The three are inextricably linked.

---

6. http://biologos.org/blog/signature-in-the-cell/

To be more specific, Meyer's response does address—as he promised—Ayala's main theological argument, namely, the argument that junk DNA shows that the human genome could not have been intelligently designed by God because it is chalk full of nonsense DNA. To refute Ayala's theological argument, Meyer shows it is based upon false scientific claims. But Falk declined to publish that part of his response until later in the week. Fair enough, but then why criticize Meyer for acting in bad faith in a preamble to the first part of his response on Monday knowing full well his response to Ayala's theological argument is coming later?

You can only appreciate theistic evolutionists for finally agreeing to engage in dialogue, but arbitrarily limiting what can be said by the other side—tying one, or both, hands behind their back—is hardly an equitable way to hold a meaningful exchange of views. Anyone who has read the reviews in question knows that only a fool would agree to the condition of totally conceding the scientific facts to Ayala, especially since his theological argument is based upon false scientific claims. Implicitly accepting Ayala's say-so on the science, "living legend" or not, would pull the legs out from under any philosophical or theological argument that Meyer chose to make. Steve Meyer, no fool, assures me that he never agreed to such a condition.

Falk is not a fool either, I assume. Neither is Ayala. So what, then, is wrong with these people?

3. Falk's Rejoinder to Meyer's Response to Ayala's "Essay" on Meyer's Book

Jay Richards

I've followed the back and forth between Francisco Ayala and Steve Meyer with interest. I happened to have just read Meyer's book *Signature in the Cell* when I first saw Ayala's commentary/review on it[7] at the BioLogos Foundation website. My initial response was that Ayala obviously hadn't read the book, and, as a result, made some embarrassing mistakes that any reader of the book would recognize.

Darrell Falk at the BioLogos Foundation was apparently responsible for inviting Ayala to comment on Meyer's book, and has been drawn into the debate.

He published the first part of Meyer's response to Ayala, but not without first offering his "background comments" about the debate. (I think David Klinghoffer has said what needs to be said about that.)[8] The BioLogos Foundation is committed to the "science-and-religion dialogue." In my opinion, however, they have a peculiar way of fostering dialogue.

BioLogos has also "updated"[9] their introduction to Ayala's "essay"—which is what they call it—to explain that Ayala wasn't invited to write a "formal review" of the book. Fair enough. But whether it's a "review," an "essay," a "response," a "commentary," or just "random thoughts," Ayala's is clearly critiquing Steve Meyer's book, *Signature in the Cell*. But his critique is clearly based on an almost complete ignorance of the book. For instance, Ayala claims: "The keystone argument of *Signature of the Cell* is that chance, by itself, cannot account for the genetic information found in the genomes

---
7. http://biologos.org/blog/on-reading-the-cells-signature/
8. http://www.evolutionnews.org/2010/03/when_theistic_evolutionists_at.html
9. http://biologos.org/blog/on-reading-the-cells-signature/

of organisms. I agree. And so does every evolutionary scientist, I presume. Why, then, spend chapter after chapter and hundreds of pages of elegant prose to argue the point?"

No one who even skimmed the book would say something this inaccurate. The inaccuracy is so blatant that I would think that Falk would be hoping that the embarrassing incident would soon be forgotten. But instead, he keeps re-opening the wound with another scratch. Now he's offered another longish commentary on Ayala's "essay" on Meyer's book, "A Rejoinder to Stephen C. Meyer's Response to Francisco Ayala."[10] And he promises that there are more to come.

Although he wisely doesn't claim that Ayala actually read Meyer's book, Falk starts by defending Ayala's claim about "hundreds of pages":

> Meyer says he only spent 55 pages on the question. By Meyer's definition of chance on page 176, and by the fact that Meyer himself refers to the competing hypotheses as "chance theories" (see pages 195, 196, and 227, for example), I happen to think that Ayala is right—it is much more than 55 pages. However, this is a side issue to what I think we should really discuss.

Hmm. So the existence of three references to chance theories in a 508-page book confirms Ayala point? Hardly. In his statement, Ayala completely misrepresents Meyer's thesis. The bit about hundreds of pages merely adds the patina of precise quantification to his misrepresentation.

Falk then raises two "concerns." The first, apparently, is that Meyer treats chance at all:

> I began my post-graduate career in genetics over four decades ago. I have taught courses such as genetics, cell biology and molecular biology for almost 35 years. I cannot recall any textbook in any course that ever seriously considered what Dr. Meyer called the "chance hypothesis." No one ever needed to do calculations of the sort that Meyer does in his book. To my recollection it was never seriously considered. Everyone knew it couldn't have worked that way.

He then goes on to say that Meyer suggests that theorists have continued to entertain the chance hypothesis for the origin of life up to the present:

---

10. http://biologos.org/blog/a-rejoinder-to-stephen-c-meyers-response-to-francisco-ayala-part-i/

Meyer seems to imply (pages 204-213) that scientists were really engaged by this hypothesis for some period of time beyond a meeting in 1966 when it was first raised. He cites work in the late 1980s and up to 2007. He seems to imply that the chance hypothesis (pure chance, from building blocks) had actually engaged origin-of-life researchers throughout this time period.

BECAUSE OF THIS alleged dismissal of chance on the part of origin-of-life researchers for the last four decades, Falk can't imagine whom Meyer has in mind as readers for his book.

A few responses to these charges:

(1) Prominent figures like Francis Crick and George Wald *did* entertain chance theories in the 1950s and 1960s. Here's Wald (quoted in one of those pesky pages in *Signature in the Cell* where Meyer talks about chance): "Time is in fact the hero of the plot.... Given so much time, the impossible becomes possible, the possible probable, and the probable virtually certain." And that's exactly what Meyer points out in the book.

(2) Pure chance ceased to be a serious contender in the 1960s as Meyer points out for the reasons that he explains in the book. He is a clear-thinking philosopher of science, interested in explaining things for the general reader who lacks detailed background knowledge, and so he lays out the arguments, the reasons, the probabilities, and the evidence painstakingly.

(3) Meyer simply does not claim that pure chance hypotheses have been leading contenders in recent decades. In fact, he quite clearly says just the opposite. On page 204, which Falk references, Meyer is talking about a conference in the 1960s. Later he talks about experimental evidence demonstrating the extreme rarity of functional sequences of amino acids—evidence that didn't exist in the 1960s—but which, as he explains, has confirmed the earlier intuitions and judgments about the insufficiency of chance by scientists in the 1960s who lacked this information.

(4) Chance nevertheless remains an important category of explanation because it continues to be a *component* in current theories such as the RNA world scenario. In fact, many current origin-of-life scenarios combine both chance and a selective mechanism as recommended by Jacques Monod's

famous book *Chance and Necessity*. Thus, Meyer's analysis of the limits of chance as a plausible explanation (or aspect of an explanation) is highly relevant to assessing many current theories of the origin of life.

(5) To make a clear and complete argument, chance needs to be treated as one of the *logical* possibilities. That's what Meyer does in his book. Why doesn't Falk get this simple point? Falk seems to think that because the community he's been swimming in hasn't bothered to reflect carefully on the full range of logically possible options, therefore it's problematic that Meyer would do so.

(6) I'm guessing that Meyer's ideal reader is the open-minded, logical person who can follow a good, clear argument, based on public evidence, and isn't intimated into mental fogginess because of social pressures not to discuss the topic of his book. Moreover, since his is a trade-press book, Meyer doesn't have the luxury of assuming that every reader will know—as apparently Falk does—why chance is so extraordinarily implausible as a complete explanation for the origin of life. So he assumes his reader will need to have that information provided in the text.

(7) That said, I'm still glad that Falk (and apparently Ayala) agree with Meyer that pure chance is not, these days, a live alternative. Unfortunately, Falk doesn't seem to realize that he is *agreeing* with Meyer on this point.

Falk's second concern is with Meyer's central positive claim. He argues that Meyer never justifies his central claim that "the activity of conscious and rational agents is the only known cause by which large amounts of new functional information arises, at least when starting from purely physical and chemical antecedents."

He attempts to refute Meyer's claim by asserting that the fact of biological evolution disproves Meyer's contention. As he explains:

> Virtually all biologists today consider it a fact that all multi-cellular organisms are derived from a single cell. Does not the information required to make the vast array of living organisms constitute Meyer's definition of "huge?" Doesn't the process of natural selection, group selection, genetic drift, and sexual selection fit his criteria of purely chemical and physical causes? There is nothing more founda-

tional to biology than that huge amounts of information has arisen through physical and chemical antecedents.

Falk cites belief in biological evolution as a counterexample of Meyer's claim. But Meyer, in the quote and in his book, is quite obviously talking about chemical evolution and the origin of life, not the evolution of life after the first reproducing cell. That's why he says: "at least when starting from *purely physical and chemical antecedents.*" In other words, Meyer, without conceding the point about biological evolution, is arguing here only about the origin of biological information from physics and chemistry—about chemical evolution—and not about what happens once you have life. And contrary to what Falk says, Meyer extensively substantiates his claim about the power of, and the need for, intelligence in producing functional information (at least, if you are starting from physical and chemical, rather than living, antecedents). The only way to fully appreciate that, however, is to read the book (especially Chapters 15 and 16 where he develops his positive case in detail).

THAT SAID, EVEN if Meyer's book were about biological evolution, Falk's argument would fall short. Falk is confusing sociology with biology. That most biologists *assume* that universal common ancestry is a fact isn't evidence for said fact. It's a fact about prominent beliefs within a community. And even if universal common ancestry is a fact, it's not evidence that all the organisms that evolved from said ancestor did so purely by a process of chance and (merely physical) necessity without the contributions of intelligence. (Oddly, Falk wants to have it both ways, since he says: "I want to be quick to add that, as a Christian, I believe that it happened at God's command and as the result of God's presence.")

In any case, that many biologists *believe* that selection and random mutation can generate large amounts of new biological information is a sociological, not a biological, fact. And frankly, it's not even a sociological fact. There are many biologists who doubt it, and get on quite well nonetheless.

4. Lying for Darwin

David Klinghoffer

Over the past couple of months at Jerry Coyne's blog, Why Evolution Is True, he and Matthew Cobb have written several blog posts attacking Stephen Meyer's *Signature in the Cell*—at this writing, five posts. The most recent by Coyne accuses Meyer of dishonesty:[11]

> Meyer does *not* mean well. He is spreading lies and confusing people by distorting real science. Is that the unfortunate result of "meaning well"? Do you think that because somebody is a "Christian brother," he's incapable of lying for Jesus?

Isn't it strange, though, that for all the persistent attacks on Meyer, in quite personal terms, Professor Coyne hasn't dared to actually read Steve's book? That's obvious because Coyne's throwaway summary of its contents—*Signature* "maintains that cells must have been designed by God because they're too complex to have evolved"—is an absurd misrepresentation. Even someone who had only read reviews of the book would know as much. Has Coyne in fact read the critical review of *Signature*, by Darrel Falk, on which he bestows approval? Or Meyer's detailed response to Falk, which Coyne dismisses as "more of the same ID pap"? Unless he's a very poor reader—and being a professor at the University of Chicago would presumably indicate otherwise—you do get the strong impression that he's commenting upon a bunch of writing by other people without having read it, certainly not with any care. Maybe he's too busy playing with his cats that he makes so much of on his blog. Or maybe he's sloppy. This is the same Dr. Coyne who earlier characterized Steve Meyer as a "young-earth creationist," which of course he's not.

But I dunno, attacking someone else for writing something that you haven't read or even carefully read about strikes me as just plain old dishon-

---

11. http://whyevolutionistrue.wordpress.com/2010/01/28/accommodationists-vs-creationists-we-all-lose/

est. If you add to that Coyne's braying slurs against Steve Meyer as "lying for Jesus," a "lying liar,"[12] etc., then to the charge of dishonesty I think you'd have to add hypocrisy as well.

---

12. http://whyevolutionistrue.wordpress.com/2009/07/22/stephen-meyer-lies-again/

5. Responding to Stephen Fletcher in the *Times Literary Supplement*

Stephen C. Meyer

Signature in the Cell stirred up debate and attracted attention as philosopher Thomas Nagel's selection of *SITC* as one of the Books of the Year brought on an interesting series of letters. Nagel was attacked (he responded, and he was attacked again) by a Darwinist who told people to forgo reading *SITC* and instead just read Wikipedia.[13] Below, Stephen Meyer himself responds in a letter, of which a shortened version was published in the *TLS*. Nagel himself responded with a letter that was published on the same page.   —Editor

To the Editor
The Times Literary Supplement
Sir—

I have been honored by the recent attention my book *Signature in the Cell* has received in your letters section following Thomas Nagel's selection of it as one of your books of the year for 2009.

Unfortunately, the letters from Stephen Fletcher criticizing Professor Nagel for his choice give no evidence of Dr. Fletcher having read the book or any evidence of his comprehending the severity of the central problem facing chemical evolutionary theories of life's origin.

In *Signature in the Cell*, I show that, in the era of modern molecular genetics, explaining the origin of the first life requires—first and foremost—explaining the origin of the information or digital code present in DNA and RNA. I also show that various theories of undirected chemical evolu-

---

13. Fletcher's letters may be accessed at http://entertainment.timesonline.co.uk/tol/arts_and_entertainment/the_tls/article6940536.ece; and at http://entertainment.timesonline.co.uk/tol/arts_and_entertainment/the_tls/tls_letters/article6959089.ece. The Meyer letter is at: http://entertainment.timesonline.co.uk/tol/arts_and_entertainment/the_tls/tls_letters/article6986702.ece.

tion—including theories of pre-biological natural selection—fail to explain the origin of the information necessary to produce the first self-replicating organism.

Yet in his letters to the *TLS* (2 and 16 December), Stephen Fletcher rebukes Nagel (and by implication my book) for failing to acknowledge that "natural selection is a chemical as well as a biological process." Fletcher further asserts that this process accounts for the origin of DNA and (presumably) the genetic information it contains.

Not only does my book address this very proposal at length, but it also demonstrates why theories of pre-biotic natural selection involving self-replicating RNA catalysts—the version of the idea that Fletcher affirms—fail to account for the origin of genetic information.

Indeed, either Dr. Fletcher is bluffing or he is himself ignorant of the many problems that this proposal faces.

First, "ribozyme engineering" experiments have failed to produce RNA replicators capable of copying more than about 10 percent of their nucleotide base sequences (Wendy K. Johnston, et al., "RNA-Catalyzed RNA Polymerization," *Science* 292 (2001): 1319-25). Yet, for natural selection to operate in an RNA World (in the strictly chemical rather than biological environment that Fletcher envisions) RNA molecules capable of fully replicating themselves must exist.

Second, everything we know about RNA catalysts, including those with partial self-copying capacity, shows that the function of these molecules depends upon the *precise* arrangement of their information-carrying constituents (i.e., their nucleotide bases). Functional RNA catalysts arise only once RNA bases are specifically arranged into information-rich sequences—that is, function arises after, not before, the information problem has been solved.

FOR THIS REASON, INVOKING PRE-BIOTIC NATURAL SELECTION IN AN RNA World does not solve the problem of the origin of genetic information; it merely *presupposes* a solution to the problem in the form of a hypothetical but necessarily *information-rich* RNA molecule capable of copying itself. As Nobel laureate Christian de Duve has noted, postulations of pre-biotic natural selection typically fail because they "need information which implies they have to presuppose what is to be explained in the first place."

Third, the capacity for even partial replication of genetic information in RNA molecules results from the activity of chemists, that is, from the *intelligence* of the "ribozyme *engineers*" who design and select the features of these (partial) RNA replicators. These experiments not only demonstrate that even highly limited forms of RNA self-replication depend upon information-rich RNA molecules, they inadvertently lend additional support to the hypothesis that intelligent design is the only known cause by which functional information arises.

Stephen C. Meyer, PhD *Cantab*
Senior Research Fellow
Discovery Institute
Seattle, Washington, USA

6. Responding Again to Stephen Fletcher in the *Times Literary Supplement*

Stephen C. Meyer

After philosopher Thomas Nagel selected Signature in the Cell *as one of 2009's best books, the* Times Literary Supplement *published a vigorous back and forth in its letters section. The final salvo[14] was by Loughborough University chemistry professor Stephen Fletcher. The response below was submitted by Stephen Meyer to* TLS, *but the editors chose not to publish it.* —Editor

To the Editor

The Times Literary Supplement

Sir—

I see that Professor Stephen Fletcher has written yet another letter (February 3, 2010) attempting to refute the thesis of my book *Signature in the Cell*. This time he cites two recent experiments in an attempt to show the plausibility of the RNA world hypothesis as an explanation for the origin of the first life. He claims these experiments have rendered the case I make for the theory of intelligent design obsolete. If anything, they have done just the reverse.

To support his claim that scientific developments have "overtaken Meyer's book," Fletcher cites, first, a scientific study by chemists Matthew Powner, Béatrice Gerland and John Sutherland of the University of Manchester (*Nature* 459, pp. 239–42). This study does partially address one, though only one, of the many outstanding difficulties associated with the RNA world scenario, the most popular current theory of the undirected chemical evolution of life. Starting with several simple chemical compounds, Powner and

---

14. Fletcher's letter may be accessed here: http://entertainment.timesonline.co.uk/tol/arts_and_entertainment/the_tls/article7013742.ece.

colleagues successfully synthesized a pyrimidine ribonucleotide, one of the building blocks of the RNA molecule.

Nevertheless, this work does nothing to address the much more acute problem of explaining how the nucleotide bases in DNA or RNA acquired their specific information-rich arrangements, which is the central topic of my book. In effect, the Powner study helps explain the origin of the "letters" in the genetic text, but not their specific arrangement into functional "words" or "sentences." Moreover, Powner and colleagues only partially addressed the problem of generating the constituent building blocks of RNA under plausible pre-biotic conditions. The problem, ironically, is their own skillful intervention. To ensure a biologically relevant outcome, they had to intervene—repeatedly and intelligently—in their experiment: first, by selecting only the right-handed isomers of sugar that life requires; second, by purifying their reaction products at each step to prevent interfering cross-reactions; and third, by following a very precise procedure in which they carefully selected the reagents and choreographed the order in which they were introduced into the reaction series. Thus, not only does this study not address the problem of getting nucleotide bases to arrange themselves into functionally specified sequences, but the extent to which it does succeed in producing biologically relevant chemical constituents of RNA actually illustrates the indispensable role of intelligence in generating such chemistry. The second study that Fletcher cites illustrates this problem even more acutely.

This work, conducted by Tracey Lincoln and Gerald Joyce (*Science* 323, pp. 229–32), ostensibly establishes the capacity of RNA to self-replicate, thereby rendering plausible one of the key steps in the RNA world scenario. Nevertheless, the "self-replicating" RNA molecules that Lincoln and Joyce construct are not capable of copying a template of genetic information from free-standing nucleotides as the protein (polymerase) machinery does in actual cells. Instead, in Lincoln and Joyce's experiment, a pre-synthesized specifically sequenced RNA molecule merely catalyzes the formation of a single chemical bond, thus fusing two other pre-synthesized partial RNA chains. Their version of "self-replication" amounts to nothing more than joining two sequence-specific pre-made halves together.

More significantly, Lincoln and Joyce themselves intelligently arranged the base sequences in these RNA chains. They generated the functionally specific information that made even this limited form of replication possible. Thus, as I argue in *Signature in the Cell*, Lincoln and Joyce's experiment not only demonstrates that even limited capacity for RNA self-replication depends upon information-rich RNA molecules, it also lends additional support to the hypothesis that intelligent design is the only known means by which functional information arises.

Stephen C. Meyer, PhD *Cantab*
Senior Research Fellow
Discovery Institute
Seattle, Washington, USA

7. Responding to Stephen Fletcher in the *Times Literary Supplement*

David Berlinski

To the Editor

The Times Literary Supplement

Sir—

Having with indignation rejected the assumption that the creation of life required an intelligent design, Mr. Fletcher has persuaded himself that it has proceeded instead by means of various chemical scenarios.[15]

These scenarios all require intelligent intervention. In his animadversions, Mr. Fletcher suggests nothing so much as a man disposed to denounce alcohol while sipping sherry.

The RNA world to which Mr. Fletcher has pledged his allegiance was introduced by Carl Woese, Leslie Orgel and Francis Crick in 1967. Mystified by the appearance in the contemporary cell of a chicken in the form of the nucleic acids, and an egg in the form of the proteins, Woese, Orgel and Crick argued that at some time in the past, the chicken was the egg.

This triumph of poultry management received support in 1981, when both Thomas Cech and Sidney Altman discovered the first of the ribonucleic enzymes. In 1986, their discoveries moved Walter Gilbert to declare the former existence of an RNA world. When Harry Noeller discovered that protein synthesis within the contemporary ribosome is catalyzed by ribosomal RNA, the existence of an ancient RNA world appeared "almost certain" to Leslie Orgel.

And to Mr. Fletcher, I imagine.

---

15. Fletcher's letters may be accessed here http://entertainment.timesonline.co.uk/tol/arts_and_entertainment/the_tls/article6940536.ece; and here http://entertainment.timesonline.co.uk/tol/arts_and_entertainment/the_tls/tls_letters/article6959089.ece.

If experiments conducted in the here and now are to shed light on the there and then, they must meet two conditions: They must demonstrate in the first place the existence of a detailed chemical pathway between RNA precursors and a form of self-replicating RNA; and they must provide in the second place a demonstration that the spontaneous appearance of this pathway is plausible under pre-biotic conditions.

The constituents of RNA are its nitrogenous bases, sugar, and phosphate. Until quite recently, no completely satisfactory synthesis of the pyrimidine nucleotides has been available.

The existence of a synthetic pathway has now been established (Matthew W. Powner et al., "The synthesis of activated pyrimidine ribonucleotides in prebiotically plausible conditions," *Nature* 459, pp. 239–242).

Questions of pre-biotic plausibility remain. Can the results of Powner et al. be reproduced without Powner et al.?

It is a question that Powner raises himself: "My ultimate goal," he has remarked, "is to get a living system (RNA) emerging from a one-pot experiment."

Let us by all means have that pot, and then we shall see further.

If the steps leading to the appearance of the pyridimines in a pre-biotic environment are not yet plausible, then neither is the appearance of a self-replicating form of RNA. Experiments conducted by Tracey Lincoln and Gerald Joyce at the Scripps Institute have demonstrated the existence of self-replicating RNA by a process of in vitro evolution. They began with what they needed and purified what they got until they got what they wanted.

Although an invigorating piece of chemistry, what is missing from their demonstration is what is missing from Powner's and that is any clear indication of pre-biotic plausibility.

I should not wish to leave this discussion without extending the hand of friendship to every party.

Mr. Nagel is correct in remarking that Mr. Fletcher is insufferable. Mr. Walton is correct in observing that the RNA world is imaginary. And Mr. Fletcher is correct in finding the hypothesis of intelligent design unacceptable.

He should give it up himself and see what happens.

David Berlinski

Paris

8. Why Are Darwinists Scared to Read *Signature in the Cell*?

David Klinghoffer

It's somehow cheering to know that while the pompous know-nothingism of Darwinian atheists in the U.S. is matched by those in England, so too not only in our country but in theirs the screechy ignorance receives its appropriate reply from people with good sense and an open mind. Some of the latter include atheists who, however, arrived at their unbelief through honest reflection rather than through the mind-numbing route of fealty to Darwinist orthodoxy. Such a person is Thomas Nagel, the distinguished NYU philosopher. He praised Stephen Meyer's *Signature in the Cell: DNA and the Evidence for Intelligent Design* in the *Times Literary Supplement* as a "book of the year," concluding with this enviable endorsement:

> [A] detailed account of the problem of how life came into existence from lifeless matter—something that had to happen before the process of biological evolution could begin.... Meyer is a Christian, but atheists, and theists who believe God never intervenes in the natural world, will be instructed by his careful presentation of this fiendishly difficult problem.

Nagel's review elicited howls from Darwinists who made no effort to pretend they had even weighed the 611-page volume in their hand, much less read a page of it. On his blog, Why Evolution Is True,[16] University of Chicago biologist Jerry Coyne complained that they hadn't ought to let such an opinion even appear in the august columns of the *TLS*:

"Detailed account"?? How about "religious speculation"?

---

16. Coyne's comments may be accessed here: http://whyevolutionistrue.wordpress.com/2009/12/01/distinguished-philosopher-blurbs-intelligent-design-book/; and here: http://whyevolutionistrue.wordpress.com/2009/12/02/more-on-nagel-and-meyer/.

Nagel is a respected philosopher who's made big contributions to several areas of philosophy, and this is inexplicable, at least to me. I have already called this to the attention of the *TLS*, just so they know.

No doubt the editors appreciated his letting them know they had erred by printing a view not in line with the official catechism. Coyne then appealed for help. Not having read the book himself, while nevertheless feeling comfortable dismissing it as "religious speculation," he pleaded:

> Do any of you know of critiques of Meyer's book written by scientists? I haven't been able to find any on the Internet, and would appreciate links.

Coyne was later relieved when a British chemist, Stephen Fletcher, published a critical letter to the editor in the *TLS* associating Meyer's argument with a belief in "gods, devils, pixies, fairies" and recommending that readers learn about chemical evolution by, instead, reading up on it elsewhere from an unimpeachable source of scientific knowledge:

> Readers who wish to know more about this topic are strongly advised to keep their hard-earned cash in their pockets, forgo Meyer's book, and simply read "RNA world" on Wikipedia.

Responding in turn with his own letter to the editor, Nagel seemed to express doubt whether the chemist had actually read *Signature in the Cell* before writing to object to Nagel's praise:

> Fletcher's statement that "It is hard to imagine a worse book" suggests that he has read it. If he has, he knows that it includes a chapter on "The RNA World" which describes that hypothesis for the origin of DNA at least as fully as the Wikipedia article that Fletcher recommends. Meyer discusses this and other proposals about the chemical precursors of DNA, and argues that they all pose similar problems about how the process could have got started.

Nagel's letter appeared beside another from a different British chemist, John C. Walton at the University of St. Andrews, who presumably did read the book since he blurbs it on the back cover as a "delightful read." In his letter, Walton reflects:

> It is an amusing irony that while castigating students of religion for believing in the supernatural, [Fletcher] offers in its place an en-

tirely imaginary "RNA world" the only support for which is speculation!

ARE YOU NOTICING a pattern here at all? All the people who hate Meyer's book appear not to have read it. So too we have the complaint of Darwinian-atheist agitator P. Z. Myers,[17] a popular blogger and biologist. Myers explains that he was unable to read the book, which he slimes as a "stinker" and as "drivel," due to his not having received a promised free review copy! But rest assured. The check is in the mail: "I suppose I'll have to read that 600 page pile of slop sometime… maybe in January."

Dr. Myers teaches at the Morris, Minnesota, satellite campus of the University of Minnesota, a college well known as the Harvard of Morris, Minnesota. So you know when he evaluates a book and calls it "slop," a book on which he has not laid on eye, that's a view that carries weight.

In all seriousness, what is this with people having any opinion at all of a book that, allow me to repeat, *they haven't read* and of which, as with Jerry Coyne, they admit they haven't so much as read a review? Even a far more measured writer like Jonathan Derbybshire, reporting for the *New Statesman* on the Nagel-*TLS* dustup, concedes, "I haven't read Myer's [sic] book, nor am I competent to assess Fletcher's contention that Nagel had simply got the science wrong." Honesty counts for something, though Derbyshire (not to be confused with *National Review*'s John Derbyshire) might have at least taken the trouble to spell Steve Meyer's name correctly.

Alas, carelessness and dishonesty are hallmarks of the Darwinian propagandists. Hordes of whom, by the way, have been trying to overwhelm *Signature*'s Amazon page. They post abusive "reviews" making, again, little pretense of having turned a single page even as they then try to boost their own phony evaluations by gathering in mobs generated by email lists and clicking on the Yes button at the question, "Was this review helpful to you?" Per Amazon's easily exploited house rules, this has the effect of boosting the "review" to enhanced prominence. It's a fraudulent tactic, and sadly typical.

---

17. http://scienceblogs.com/pharyngula/2009/12/you_know_its_a_stinker_when_th.php

9. Every Bit Digital: DNA's Programming Really Bugs Some ID Critics

Casey Luskin

GOOGLE'S CORPORATE MOTTO IS "DON'T BE EVIL," BUT UNFORTUnately, not all who work at the search engine behemoth seem to practice the slogan. Mark Chu-Carroll, a mathematician and Google software engineer, called Stephen Meyer's *Signature in the Cell* "a rehash of the same old s—t," even though he admitted, "I have not read any part of Meyer's book." Chu-Carroll further decried the "dishonesty" of Meyer, whom he called a "bozo" for merely claiming that DNA contains "digital code" that functions like a "computer."

It seems that Meyer's book isn't the only relevant literature that Chu-Carroll hasn't read.

In 2003 renowned biologist Leroy Hood and biotech guru David Galas authored a review article in the world's leading scientific journal, *Nature*, titled, "The Digital Code of DNA." The article explained, "A remarkable feature of the structure is that DNA can accommodate almost any sequence of base pairs—any combination of the bases adenine (A), cytosine (C), guanine (G) and thymine (T)—and, hence any digital message or information." MIT Professor of Mechanical Engineering Seth Lloyd (no friend of ID) likewise eloquently explains why DNA has a "digital" nature:

> It's been known since the structure of DNA was elucidated that DNA is very digital. There are four possible base pairs per site, two bits per site, three and a half billion sites, seven billion bits of information in the human DNA. There's a very recognizable digital code of the kind that electrical engineers rediscovered in the 1950s that maps the codes for sequences of DNA onto expressions of proteins.

DNA's computer-like attributes have also been noted by leading thinkers. Software mogul Bill Gates said, "Human DNA is like a computer program but far, far more advanced than any software we've ever created." Francis Collins—who headed the Human Genome project and is a noted proponent of Darwinism, describes DNA as a "digital code," and observes that "DNA is something like the hard drive on your computer" that contains "programming." Even Richard Dawkins has observed that "the machine code of the genes is uncannily computer-like. Apart from differences in jargon, the pages of a molecular biology journal might be interchanged with those of a computer engineering journal."

But what is the computer code doing? It turns out that it's programming nothing less than nanotechnology—micromolecular machines inside the cell. In the words of Bruce Alberts, former president of the U.S. National Academy of Sciences, "The entire cell can be viewed as a factory that contains an elaborate network of interlocking assembly lines, each of which is composed of a set of large protein machines."

For Chu-Carroll to ignore the many leading evolutionary scientists and thinkers who have compared the cell to computers or machines, and instead to accuse Meyer of "dishonesty" is, well, a low form of argument that the Google motto probably prevents us from naming. But where in our experience do digital code, computer programming, and factories filled with machines come from? Chu-Carroll knows the answer, which is probably why he doesn't like Meyer's argument.

10. Ayala: "For the Record, I Read *Signature in the Cell*"

Jay Richards

Over at BioLogos, Professor Francisco Ayala has responded to *Signature of Controversy*—the collection of responses to criticisms of *Signature in the Cell*. As with the previous Ayala response at BioLogos, this one includes an introduction by Darrell Falk.

The burden of Ayala's response is to wax indignant that some of us have suggested, based on his original "response" to *Signature in the Cell*, that he had not actually read the book. Why would we suggest that? Well, because he so profoundly misrepresented Meyer's thesis.

Here's what he said: "The keystone argument of *Signature* [sic] *of the Cell* is that chance, by itself, cannot account for the genetic information found in the genomes of organisms." He goes on to suggest that Meyer spends "most" of his book attempting to refute the chance hypothesis. Really.

This is such a whopper that I would have expected Ayala not to bring it up again. But in his current response, he begins:

> Dr. Stephen Meyer writes: "eminent evolutionary biologist Francisco Ayala does not appear to have even made a search for the crib notes online. Indeed... it appears that he did little more than glance at the title page and table of contents" (p. 9). David Klinghoffer disagrees: "My colleague Dr. Meyer thinks Ayala did read the Table of Contents, but I must disagree" (p. 19).
>
> Is this the kind of language Meyer and Klinghoffer want to use to engage in constructive dialogue with their critics? Or does it represent a distinctive way in which members of the Discovery Institute seek to practice Christian charity?
>
> For the record, I read *Signature in the Cell*.

To justify his original characterization of Meyer's book, Ayala then offers an analysis of the index of *Signature in the Cell*, which lists a variety of pages in which "chance" appears, to establish his original assertion. This discussion is beginning to enter the Twilight Zone.

Ayala implies that Discovery Institute folks have failed to practice Christian charity in suggesting that he didn't read *Signature in the Cell* before commenting on it. (This is a curious complaint, coming from Ayala, who has repeatedly charged that intelligent design is blasphemy.)[18] But does anyone—friend or critic—actually believe that Ayala provided an accurate summary of Meyer's argument? Would Darrell Falk, for instance, be willing to state, on the record, that "[t]he keystone argument of Signature [in] the Cell is that chance, by itself, cannot account for the genetic information found in the genomes of organisms"?

I'M WILLING TO state, for the record, that I don't think that any competent reader could read *SITC*, with its chapters lovingly devoted to critiques of self-organizational theories, pre-biotic selection theories that combine chance and necessity, and which explores the possibility of chance while clearly stating that the chance hypothesis is no longer seriously entertained by origin-of-life theorists, and come away thinking that Ayala had accurately summarized the book. That would be true even if the word "chance" occurred on every single page. You don't find the thesis of a book in its index, for goodness' sake.

Given the above assumption, and the assumption that Ayala is a competent reader—both ideas shared by my colleagues who have weighed in on this question—by far *the most charitable interpretation* of Ayala's summary is that he didn't read the book. If he really has read *Signature in the Cell*, and yet is still defending his false summary, what should we conclude?

---

18. http://www.examiner.com/x-36629-Hartford-Freethought-Examiner~y2010m3d29-The-Idea-of-Intelligent-Design-is-Blasphemous

II

ON READING STEPHEN MEYER'S *SIGNATURE IN THE CELL*

11. Responding to Darrel Falk's Review of *Signature in the Cell*

Stephen C. Meyer

Dr. Darrel Falk, biology professor at Point Loma Nazarene University, reviewed *Signature in the Cell* for the BioLogos Foundation's blog, "Science & the Sacred."[19] Below is Dr. Meyer's response. —Editor

In 1985 I attended a conference that brought a fascinating problem in origin-of-life biology to my attention—the problem of explaining how the information necessary to produce the first living cell arose. At the time, I was working as a geophysicist doing digital signal processing, a form of information analysis and technology. A year later, I enrolled in graduate school at the University of Cambridge, where I eventually completed a PhD in the philosophy of science after doing interdisciplinary research on the scientific and methodological issues in origin-of-life biology. In the ensuing years, I continued to study the problem of the origin of life and have authored peer-reviewed and peer-edited scientific articles on the topic of biological origins, as well as co-authoring a peer-reviewed biology textbook. Last year, after having researched the subject for more than two decades, I published *Signature in the Cell*, which provides an extensive evaluation of the principal competing theories of the origin of biological information and the related question of the origin of life. Since its completion, the book has been endorsed by prominent scientists including Philip Skell, a member of the National Academy of Sciences; Scott Turner, an evolutionary biologist at the State University of New York; and Professor Norman Nevin, one of Britain's leading geneticists.

---

19. http://biologos.org/blog/signature-in-the-cell/

Nevertheless, in his recent review on the BioLogos website, Professor Darrel Falk characterizes me as merely a well-meaning, but ultimately unqualified, philosopher and religious believer who lacks the scientific expertise to evaluate origin-of-life research and who, in any case, has overlooked the promise of recent pre-biotic simulation experiments. On the basis of two such experiments, Falk suggests I have jumped prematurely to the conclusion that pre-biotic chemistry cannot account for the origin of life. Yet neither of the scientific experiments he cites provides evidence that refutes the argument of my book or solves the central mystery that it addresses. Indeed, both experiments actually reinforce—if inadvertently—the main argument of *Signature in the Cell*.

The central argument of my book is that intelligent design—the activity of a conscious and rational deliberative agent—best explains the origin of the information necessary to produce the first living cell. I argue this because of two things that we know from our uniform and repeated experience, which following Charles Darwin I take to be the basis of all scientific reasoning about the past. First, intelligent agents have demonstrated the capacity to produce large amounts of functionally specified information (especially in a digital form). Second, no undirected chemical process has demonstrated this power. Hence, intelligent design provides the best—most causally adequate—explanation for the origin of the information necessary to produce the first life from simpler non-living chemicals. In other words, intelligent design is the only explanation that cites a cause known to have the capacity to produce the key effect in question.

Nowhere in his review does Falk refute this claim or provide another explanation for the origin of biological information. In order to do so, Falk would need to show that some undirected material cause has demonstrated the power to produce functional biological information apart from the guidance or activity of a designing mind. Neither Falk, nor anyone working in origin-of-life biology, has succeeded in doing this. Thus, Falk opts instead to make a mainly personal and procedural argument against my book by dismissing me as unqualified and insisting that it is "premature" to draw any negative conclusions about the adequacy of undirected chemical processes.

11. Responding to Darrel Falk's Review of Signature in the Cell

To support his claim that I rushed to judgment, Falk first cites a scientific study published last spring after my book was in press. The paper, authored by University of Manchester chemist John Sutherland and two colleagues, does partially address one of the many outstanding difficulties associated with the RNA world, the most popular current theory about the origin of the first life.

Starting with a 3-carbon sugar (D-gylceraldehyde), and another molecule called 2-aminooxazole, Sutherland successfully synthesized a 5-carbon sugar in association with a base and a phosphate group. In other words, he produced a ribonucleotide. The scientific press justifiably heralded this as a breakthrough in pre-biotic chemistry because previously chemists had thought (as I noted in my book) that the conditions under which ribose and bases could be synthesized were starkly incompatible with each other.

Nevertheless, Sutherland's work does not refute the central argument of my book, nor does it support the claim that it is premature to conclude that only intelligent agents have demonstrated the power to produce functionally specified information. If anything, it illustrates the reverse.

In Chapter 14 of my book I describe and critique the RNA world scenario. There I describe five major problems associated with the theory. Sutherland's work only partially addresses the first and least severe of these difficulties: the problem of generating the constituent building blocks or monomers in plausible pre-biotic conditions. It does not address the more severe problem of explaining how the bases in nucleic acids (either DNA or RNA) acquired their specific information-rich arrangements. In other words, Sutherland's experiment helps explain the origin of the "letters" in the genetic text, but not their specific arrangement into functional "words" or "sentences."

Even so, Sutherland's work lacks pre-biotic plausibility and does so in three ways that actually underscore my argument.

First, Sutherland chose to begin his reaction with only the right-handed isomer of the 3-carbon sugars he needed to initiate his reaction sequence. Why? Because he knew that otherwise the likely result would have had little biological significance. Had Sutherland chosen to use a far more plausible racemic mixture of both right- and left-handed sugar isomers, his reaction would have generated undesirable mixtures of stereoisomers—mixtures that would seriously complicate any subsequent biologically relevant polymerization. Thus, he himself solved the so-called chirality problem in origin-of-life chemistry by intelligently selecting a single enantiomer, i.e., only the right-handed sugars that life itself requires. Yet there is no demonstrated source for such non-racemic mixture of sugars in any plausible pre-biotic environment.

Second, the reaction that Sutherland used to produce ribonucleotides involved numerous separate chemical steps. At each intermediate stage in his multi-step reaction sequence, Sutherland himself intervened to purify the chemical by-products of the previous step by removing undesirable side products. In so doing, he prevented—by his own will, intellect and experimental technique—the occurrence of interfering cross-reactions, the scourge of the pre-biotic chemist.

Third, in order to produce the desired chemical product—ribonucleotides—Sutherland followed a very precise "recipe" or procedure in which he carefully selected the reagents and choreographed the order in which they were introduced into the reaction series, just as he also selected which side products to be removed and when. Such recipes, and the actions of chemists who follow them, represent what the late Hungarian physical chemist Michael Polanyi called "profoundly informative intervention[s]." Information is being added to the chemical system as the result of the deliberative actions—the intelligent design—of the chemist himself.

IN SUM, NOT only did Sutherland's experiment *not* address the more fundamental problem of getting the nucleotide bases to arrange themselves into functionally specified sequences; the extent to which it did succeed in producing more life-friendly chemical constituents actually illustrates the indispensable role of intelligence in generating such chemistry.

The second experiment that Falk cites to refute my book illustrates this problem even more acutely. This experiment is reported in a scientific paper[20] by Tracey Lincoln and Gerald Joyce ostensibly establishing the capacity of RNA to self-replicate, thereby rendering plausible one of the key steps in the RNA world hypothesis. Falk incorrectly intimates that I did not discuss this experiment in my book. In fact, I do on page 537.

In any case, it is Falk who draws exactly the wrong conclusion from this paper. The central problem facing origin-of-life researchers is neither the synthesis of pre-biotic building blocks (which Sutherland's work addresses) or even the synthesis of a self-replicating RNA molecule (the plausibility of which Joyce and Tracey's work seeks to establish, albeit unsuccessfully: see below).[21] Instead, the fundamental problem is getting the chemical building blocks to arrange themselves into the large information-bearing molecules (whether DNA or RNA). As I show in *Signature in the Cell*, even the extremely limited capacity for RNA self-replication that has been demonstrated depends critically on the specificity of the arrangement of nucleotide bases—that is, upon pre-existing sequence-specific information.

The Lincoln and Joyce experiment that Falk describes approvingly does not solve this problem, at least not apart from the intelligence of Lincoln and Joyce. In the first place, the "self-replicating" RNA molecules that they construct are not capable of copying a template of genetic information from free-standing chemical subunits as the polymerase machinery does in actual cells. Instead, in Lincoln and Joyce's experiment, a pre-synthesized *specifically sequenced* RNA molecule merely catalyzes the formation of a single chemical bond, thus fusing two other pre-synthesized partial RNA chains. In other words, their version of "self-replication" amounts to nothing more than joining two sequence specific pre-made halves together. More significantly, Lincoln and Joyce themselves *intelligently arranged* the matching base sequences in these RNA chains. They did the work of replication. They generated the functionally specific information that made even this limited form of replication possible.

The Lincoln and Joyce experiment actually confirms three related claims that I make in *Signature in the Cell*. First, it demonstrates that even the capac-

20. http://www.sciencemag.org/cgi/content/abstract/1167856
21. http://biologicinstitute.org/2009/04/01/biologic-institute-announces-first-self-replicating-motor-vehicle

ity for modest partial self-replication in RNA itself depends upon sequence specific (i.e., information-rich) base sequences in these molecules. Second, it shows that even the capacity for partial replication of genetic information in RNA molecules results from the activity of chemists, that is, from the intelligence of the "ribozyme engineers" who design and select the features of these (partial) RNA replicators. Third, pre-biotic simulation experiments themselves confirm what we know from ordinary experience, namely, that intelligent design is the only known means by which functionally specified information arises.

For nearly sixty years origin-of-life researchers have attempted to use pre-biotic simulation experiments to find a plausible pathway by which life might have arisen from simpler non-living chemicals, thereby providing support for chemical evolutionary theory. While these experiments have occasionally yielded interesting insights about the conditions under which certain reactions will or won't produce the various small molecule constituents of larger bio-macromolecules, they have shed no light on how the information in these larger macromolecules (particularly in DNA and RNA) could have arisen. Nor should this be surprising in light of what we have long known about the chemical structures of DNA and RNA. As I show in *Signature in the Cell*, the chemical structures of DNA and RNA allow them to store information precisely because chemical affinities between their smaller molecular subunits do not determine the specific arrangements of the bases in the DNA and RNA molecules. Instead, the same type of chemical bond (an N-glycosidic bond) forms between the backbone and each one of the four bases, allowing any one of the bases to attach at any site along the backbone, in turn allowing an innumerable variety of different sequences. This chemical indeterminacy is precisely what permits DNA and RNA to function as information carriers. It also dooms attempts to account for the origin of the information—the precise sequencing of the bases—in these molecules as the result of deterministic chemical interactions.

Nevertheless, for Professor Falk, drawing any negative conclusions about the adequacy of purely undirected chemical processes or —worse—making an inference to intelligent design, is inherently premature. Indeed, for him such thinking constitutes giving up on science or making "an argument from ignorance." But this betrays a misunderstanding of both science and the basis of the design argument that I am making.

Scientific investigations not only tell us what nature does, they also frequently tell us what nature doesn't do. The conservation laws in thermodynamics, for example, proscribe certain outcomes. The first law tells us that energy is never created or destroyed. The second tells us that the entropy of a closed system will *never* decrease over time. Moreover, because these laws are based upon our uniform and repeated experience, we have great confidence in them. That is why physicists don't, for example, still consider research on perpetual motion machines to be worth investigating or funding.

In the same way, we now have a wealth of experience showing that what I call *specified* or *functional* information (especially if encoded in digital form) does not arise from purely physical or chemical antecedents. Indeed, the ribozyme engineering and pre-biotic simulation experiments that Professor Falk commends to my attention actually lend additional inductive support to this generalization. On the other hand, we do know of a cause—a type of cause—that has demonstrated the power to produce functionally specified information. That cause is intelligence or conscious rational deliberation. As the pioneering information theorist Henry Quastler once observed, "The creation of information is habitually associated with conscious activity." And, of course, he was right. Whenever we find information—whether embedded in a radio signal, carved in a stone monument, written in a book or etched on a magnetic disc—and we trace it back to its source, invariably we come to mind, not merely a material process. Thus, the discovery of functionally specified, digitally encoded information along the spine of DNA provides compelling positive evidence of the activity of a prior designing intelligence. This conclusion is not based upon what we *don't* know. It is based upon what we *do know* from our uniform experience about the cause and effect structure of the world—specifically, what we know about what does, and does not, have the power to produce large amounts of specified information.

THAT PROFESSOR FALK rejects this knowledge *as knowledge*, and the case for design based on it, reflects his own commitment to finding a solution to the origin of life problem within a strictly materialistic framework. Indeed, he and his colleagues at *BioLogos* have made clear that they accept the principle of methodological naturalism, the idea that scientists, to be scientists, must limit themselves to positing only materialistic explanations for all phenomena. Of course, it is their right to accept this intellectual limitation on theorizing if they wish. But it needs to be noted that the principle of methodological naturalism is an arbitrary philosophical assumption, not a principle that can be established or justified by scientific observation itself. Others of us, having long ago seen the pattern in pre-biotic simulation experiments, to say nothing of the clear testimony of thousands of years of human experience, have decided to move on. We see in the information-rich structure of life a clear indicator of intelligent activity and have begun to investigate living systems accordingly. If, by Professor Falk's definition, that makes us philosophers rather than scientists, then so be it. But I suspect that the shoe is now, instead, firmly on the other foot.

12. Asking Darrel Falk to Pick a Number, Any Number

Richard Sternberg

I HAVE LONG QUESTIONED THE ASSUMPTION THAT MOST GENOMIC DNA sequences are "nonsensical" or "junk." And given the data that have emerged over the past seven or so years, a functionalist view of the genome has robust empirical support. It is for this reason that I think many of the arguments presented by the BioLogos Foundation are "wrong on many counts," to borrow a phrase from Darrel Falk.

Here is an example. While reading the "critique" of Steve Meyer's book, *Signature in the Cell*, by Francisco Ayala,[22] a number struck me that I know to be incorrect. The integer that I am referring to is "25,000" and it is claimed to be the known tally of genes in our chromosomes:

> The human genome includes about 25,000 genes and lots of other (mostly short) switch sequences...

Now, the problem with such a statement is this: While there are ~25,000 *protein-coding genes* in our DNA, the number of *RNA-coding genes* is predicted to be much higher, >450,000.[23] Some of the latter range in length from being quite short—only 20 or so genetic letters—to being millions of letters long. Since 2004 we have learned that over 90 percent of our DNA is transcribed into RNA sequences at some developmental stage, in different cell and tissue types. (Our brain cells are unusually rich in these non-translated RNAs.) These RNAs are then processed into regulatory and structural sequences of all sizes.[24]

22. http://biologos.org/blog/on-reading-the-cells-signature/
23. Rederstorff, M., S. H. Bernhart, A. Tanzer, M. Zywicki, K. Perfler, M. Lukasser, I. L. Hofacker and A. Hüttenhofer (in press), "RNPomics: Defining the ncRNA transcriptome by cDNA library generation from ribonucleo-protein particles," *Nucleic Acids Research* (2010).
24. Amaral, P. P., M.E. Dinger, T. R. Mercer and J. S. Mattick, "The eukaryotic genome as an RNA machine," *Science* 319:5871 (2008), pp. 1787–1789; Dinger, M. E., P. P. Amaral, T. R. Mercer and J. S. Mattick, "Pervasive transcription of the eukaryotic genome: functional indices and conceptual implications," *Briefings in Functional Genomics and Proteomics* 8:6 (2009), pp.

It could of course be argued, as it has been, that most of these RNA transcripts are themselves junk. But a host of them are packaged into complexes with different proteins.

So the true number of genes in our DNA is probably >450,000 + 25,000 = >475,000. What is more, these >450,000 genes cover more than 88.5 percent of our 3 billion genetic letters. That's right—most, if not close to all, of our chromosomal DNA consists of different types of genes that have only recently been discovered. How do these facts square with this comment made by Falk?[25]

> but this still doesn't negate the fact that almost certainly much, if not most, of the DNA plays no role, and in many cases can be harmful.

WELL, IT ALL depends on how he is using the words "much" and "most." I really don't know. So I have a question for him: *Exactly what fraction of the transcribed 88.5 percent of our DNA are you willing to say "plays no role" or can be harmful?* All I am asking for is a prediction, such as "90 percent of these DNA letters is superfluous" ("or 79.5 percent of the RNAs are nonsensical"). Since he also said "almost certainly" in the above statement, he must have a figure in mind. So I say pick a number, any number.... But to be a good sport, I'll show my prediction: All of the expressed 88.5 percent of our DNA has diverse roles in our development.

407–423; Mercer, T. R., I. A. Qureshi, S. Gokhan, M. E. Dinger, G. Li, J. S. Mattick and M. F. Mehler, "Long noncoding RNAs in neuronal-glial fate specification and oligodendrocyte lineage maturation," *BMC Neuroscience* 11:1 (2010), p. 14.

25. http://biologos.org/blog/a-rejoinder-to-meyer-2

13. Ayala and Falk Miss the Signs in the Genome

Richard Sternberg

In his recent response to Stephen Meyer's *Signature in the Cell*, Francisco Ayala claimed that repetitive portions of our DNA called "*Alu*" sequences are "nonsensical." Ayala wrote: "Would a function ever be found for these one million nearly identical *Alu* sequences? It seems most unlikely." In his response to Ayala, Meyer showed that Ayala is factually wrong about this. According to recent technical papers in genomics, *Alu* sequences perform multiple functions.

In a rejoinder to Meyer, Darrel Falk defended Ayala and claimed although "a number of functional regions have been discovered within *Alu* sequences," there "is no question that many *Alu* sequences really have no function."[26]

In my previous chapter, I showed that the vast majority of the genome is transcribed, either into protein-coding genes or into regulatory RNAs. The technical literature—some of which I cited in that blog—reports that the genome is an RNA-coding machine. Clearly, most DNA really *does* have function.

In this and subsequent chapters, I will provide other sorts of evidence that so-called "junk DNA" is not junk at all, but functional.

We have all seen a variant of the plot in a movie. A strange signal appears—in one film it is a recurrent wireless telegraph code that is transmitted from San Diego after a global nuclear holocaust (*On the Beach*); in another it is radio transmissions from deep space (*Contact*); in still another it is crop circles (*Signs*). In the first movie, the signal turns out to be due to a Coca-Cola bottle: Wind blowing on a window shade next to the bottle results in the

---

26. Ayala and Falk's comments may be accessed here: http://biologos.org/blog/on-reading-the-cells-signature/; and here: http://biologos.org/blog/a-rejoinder-to-meyer-2.

latter being occasionally nudged, which sometimes leads to a telegraph key being tapped by the very same. But in the second movie, the signals received turn out to contain a complex set of encrypted data with an intricate mathematical pattern—they are the specifications for building a device that can travel through space-time wormholes, sent from a friendly alien civilization. So also are the crop circles in the third film messages from an extraterrestrial race, except that the designs portend an attack on humanity.

Now, the reason we are drawn in by such stories is obvious: The signals have serious implications for the characters. It could mean the survival of mankind after a thermonuclear war; it could mean that there are other sentient beings in the universe. That is why we would quickly lose interest in the plot if, say, in every scene where a scientist appeared before an important governmental group and said, "The outer space signal contains over sixty thousand, multidimensional pages of complex architectural plans," she were countered with, "This is exactly the predicted outcome of billions of years of cosmic evolution—you see, random interstellar events lead to just this kind of complex specified information… we are not impressed." We would want our money back.

MY PURPOSE IN bringing up this subject is that I have a mysterious genomic signal for you to see—which I will show you in the next chapter. We detected it some time ago and it has aroused the interest of some genomicists, but you will find no mention of it in books such as Francis Collins's *The Language of God*—which is peculiar. But I have another aim in mind, too, in broaching this possible chromosomal code: A key first indicator of functionality is a distinctly non-random pattern. The persistence of a distinct signal in different contexts often suggests functional constraints are operative—that is why genomicists look for them. And since I want to focus on the *global* functions of such *Short Interspersed Nuclear Elements (SINEs)* as human *Alus* and their mouse and rat counterparts, their far-from-random placement cannot be elided. In fact, I will argue that it is a critical part of the genome story that the folks at BioLogos aren't telling you.

To prepare for the mysterious genomic signal, though, I want to draw your attention to this figure:

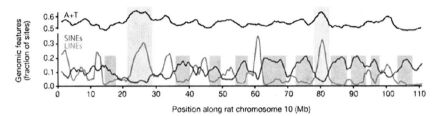

Reprinted by permission from Macmillan Publishers Ltd: figure 9d, "Genome sequence of the Brown Norway rat yields insights into mammalian evolution," *Nature* 428 (2004), pp. 493-521, copyright 2004.[27]

What you are seeing are the relative densities of *Long Interspersed Nuclear Element (LINE) L1s* and SINEs along 110,000,000 DNA letters of rat chromosome 10.[28] (From Fig. 9d of reference 1.) The x-axis represents the sequence of letters in DNA and of the two lower lines, the one that starts below the other indicates where SINEs occur—what Ayala calls "obnoxious sequences" that are supposedly due to "degenerative biological processes that are not the result of ID." The line that starts above that indicates where LINE sequences occur.

By the way, Francis Collins is a principal author of the *Nature* paper where these results are published.

Both LINEs and SINEs are types of mobile DNA, namely, *retrotransposons*, and together they can comprise around half of the mammalian genome. As is clear from the figure, especially as it was originally published in color, LINEs tend to peak in abundance where SINEs taper off and *vice versa*. We have known about this pattern since the late 1980s, so it is no surprise to someone who has been following the subject. What should be surprising to anyone, however, is that the same machinery is responsible for the movement of both types of retrotransposon. A complete L1 element encodes the pro-

---

27. http://www.nature.com/nature/journal/v428/n6982/full/nature02426.html
28. Rat Genome Sequencing Project Consortium, "Genome sequence of the Brown Norway rat yields insights into mammalian evolution," *Nature* 428 (2004), pp. 493–521.

teins necessary to "reverse transcribe" an RNA copy of itself back into DNA, and to insert the generated duplicate into some chromosomal site. SINEs, by way of contrast, rely on the L1-specified proteins for all their copying and pasting routines.

This compartmentalization of LINEs and SINEs along the mammalian chromosome can also be detected by using molecular probes for L1 or *Alu*(-like) sequences.[29]

FOR JUNKETY-JUNK ELEMENTS that can make up 50 percent of a mammal's mostly junkety-junk genome, the rule seems to be: Location, location, location.

Interestingly, this higher-order pattern cannot be detected when small sections of DNA are examined. It only becomes evident when stretches that are millions of nucleotides long are studied.

This banding pattern has been known for decades—but for some reason it is rarely (if ever) discussed by "junk DNA" advocates. The bands on the chromosome arms fall into two general categories:

- R bands: DNA compartments that are enriched with the genetic letters G and C, have a high concentration of protein-coding genes, a preponderance of Alu or Alu-like repetitive elements (e.g., mouse B1s), and replicate early in the DNA synthesis phase of the cell cycle.
- G bands: DNA compartments that are enriched with the genetic letters A and T, have a low concentration of protein-coding genes, a high density of the L1 retrotransposon, and replicate late in the DNA synthesis phase of the cell cycle.

There are other strong functional correlations, too, such as the distribution of types of chromatin and how the genome is packaged in the nucleus. But I'm getting way ahead of myself.

---

29. Chen, T. L. and L. Manuelidis, "SINEs and LINEs cluster in distinct DNA fragments of Giemsa band size," *Chromosoma* 98 (1989), pp. 309–316; Korenberg, J. R. and M. C. Rykowski, "Human genome organization: *Alu*, lines, and the molecular structure of metaphase chromosome bands," *Cell* 53:3 (1988), pp. 391–400; Costantini, M., O. Clay, C. Federico, S. Saccone, F. Auletta and G. Bernardi, "Human chromosomal bands: nested structure, high-definition map and molecular basis," *Chromosoma* 116 (2007), pp. 29–40.

Now some questions. Which DNA regions of the mammalian genome are enriched in the codes for the most essential functions? *Precisely where you find Alus and Alu-like sequences.* Which sections of the mammalian genome have the highest rates of transcription? *Precisely where you find Alus and Alu-like sequences.* Where do you find the strongest organizational correlations between any two mammalian genomes? *Precisely where you find Alus and Alu-like sequences.*

Aren't these correlations a bit strange for genomes that supposedly consist mostly of junk and are constantly being corrupted by "degenerative processes"? Why do such "obnoxious sequences" have any kind of *conserved* higher-order "bar code" pattern? These facts of mammalian chromosome biology have been known for years, if not decades. But, alas, no mention of them is to be found in the literature that wants to emphasize the unintelligent design of our genome. To make up for this lack, then, I am going to discuss such facts in more detail *after* I show you the mystery signal in the next chapter.

14. Discovering Signs in the Genome by Thinking Outside the BioLogos Box

Richard Sternberg

Having promised that I would show you a mysterious genomic signal, I shall now fulfill that promise. The previous chapter was devoted to describing the linear distribution of LINEs and SINEs along mammalian chromosomal DNA. We saw that *L1* retrotransposons tend to be densest in the regions where *Alus* and *Alu*-like elements are the least common and *vice versa*. I included the following figure from an article co-authored by Francis Collins that showed this compartmentalization of LINEs and SINEs along over a hundred million genetic letters of rat chromosome 10:

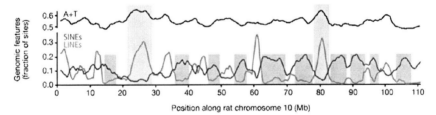

Reprinted by permission from Macmillan Publishers Ltd: figure 9d, "Genome sequence of the Brown Norway rat yields insights into mammalian evolution," *Nature* 428 (2004), pp. 493-521, copyright 2004.[30]

The line that starts lower than the other two indicates the distribution of SINEs along a 110-million base pair interval of rat chromosome 10.

Intriguing as this non-random distribution of repetitive elements may be, it gets even more interesting when one realizes that SINEs are specific to taxonomic groups. Each primate genome has distinct subfamilies of the *Alu* sequence. The mouse genome, on the other hand, has no *Alus* but it does

---

30. http://www.nature.com/nature/journal/v428/n6982/full/nature02426.html

have three unique SINE families called B1, B2, and B4. While mouse B1 shares some sequence similarity with *Alu*, it has no relationship to the B2 or B4 elements; the latter two are also unrelated to each other. What then about the rat SINEs along chromosome 10? Well, the genome of the rat has one main SINE family called ID, for the "Identifier" sequence. The ID elements have nothing in common at the DNA sequence level with the mouse B1s, B2s, or B4s, and they are wholly dissimilar to *Alu*s.

So we have three different mammal genomes (primate, mouse, and rat) and three different sets of SINEs. But since I showed you rat chromosome 10 yesterday, let's just focus on the two rodent genomes.

Now, the mouse and rat are estimated to have diverged 22 million years ago. During that interval, individual SINEs have been coming and going and going and coming, in and out of chromosomes. This ongoing insertion/deletion of these retrotransposons is precisely the "degenerative process" that Francisco Ayala referred to when mentioning *Alu*s.

For the 22 million years that have occurred since the mouse and rat lineages went their separate ways, both genomes have been subjected to hundreds of thousands—if not millions—of separate SINE insertion events. Putting on our "junk DNA" thinking caps, let's try to predict what the outcomes of such long-term mutational bombardments would be *vis-à-vis* the linear distributions of SINEs along a chromosome. To do this, let's connect these two statements:

1) "… almost certainly much, if not most, of the DNA plays no role…"

2) "Perhaps one could attribute the obnoxious presence of the *Alu* sequences to degenerative biological processes…"[31]

Or to restate, we have "much, if not most" rodent DNA that is not functional having being subjected to extensive degenerative events over the course of twenty-two million years. The only difference that we must keep in mind is that the "obnoxious" elements that were involved in this example of decay in the mouse genome are B1s, B2s, and B4s; whereas the destructive force in the rat genome in this case was primarily the ID elements.

---

31. The first is from http://biologos.org/blog/a-rejoinder-to-meyer-2 and the second from http://biologos.org/blog/on-reading-the-cells-signature/.

14. Discovering Signs in the Genome by Thinking Outside the BioLogos Box   79

Okay. What do we expect in general from degenerating processes that have no functional consequences? Let's do a thought experiment. Consider the surfaces of two moons that were once part of the same planetary body 22 million years ago. Since their separation, both have been subjected to independent collisions with asteroids, meteorites, and other pieces of space debris. Question: Would you expect the scar patterns on both to be different or identical? (It may seem like a silly question, but bear with me.)

Replace now the word "moons" with the "mouse and rat genomes" and "asteroids and meteorites and other pieces of space debris" with SINEs, and you will see what I am asking. So I'll rephrase my question. What should we expect regarding the linear distribution of independent SINE impacts along mouse and rat chromosomes?

    A. Completely independent patterns—like meteorite impact sites on moons;

    B. A few overlapping patterns, due to chance; or

    C. Nearly identical patterns.

And the mystery signal is…

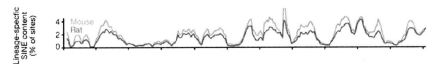

Reprinted by permission from Macmillan Publishers Ltd: figure 9c, "Genome sequence of the Brown Norway rat yields insights into mammalian evolution," *Nature* 428 (2004), pp. 493-521, copyright 2004.[32]

This is a second figure from the article co-authored by Francis Collins (from Fig. 9c of Ref. 1). The scale on the x-axis is the same as that of the previous graph—it is the same 110,000,000 genetic letters of rat chromosome 10. The scale on the y-axis is different, with the slightly lower line in this figure corresponding to the distribution of rat-specific SINEs in the rat genome (i.e., ID sequences). The *higher* line in this figure, however, corresponds to the pattern of B1s, B2s, and B4s in the *mouse* genome.

Was it what you expected from a degenerative process? Why?

---

32. http://www.nature.com/nature/journal/v428/n6982/full/nature02426.html

At this point the theistic evolutionist might say—Silly Rick: Common descent explains this pattern!

Wrong, wrong, wrong.

Let me repeat—*each graph denotes only lineage-specific mutational insertions.*

The mutational signal from mouse B1s, B2s, and B4s is equivalent to the mutational signal of rat IDs.

It almost looks as if, say, the rat graph was copied, slightly redrawn, labeled "mouse," and then pasted above the previous line. (Of course, it wasn't.) How strange that two independently acting degenerative processes—affecting mostly "junk DNA"— would lead to the same higher-order pattern.

It's a bizarre pattern. And this correlation occurs throughout both genomes.

The Rat Genome Consortium—and thus Francis Collins—apparently thought it worthy to devote a whole section to the phenomenon. Titled "Co-Localization of SINEs in Rat and Mouse," the section states:

> Despite the different fates of SINE families, the number of SINEs inserted after speciation in each lineage is remarkably similar: ~300,000 copies… Figure 9c displays the lineage-specific SINE densities on rat chromosome 10 and in the mouse orthologous blocks, *showing a stronger correlation than any other feature. The cause of the unusual distribution patterns of SINEs, accumulating in gene-rich regions* where other interspersed repeats are scarce, *is apparently a conserved feature, independent of the primary sequence of the SINE* and effective over regions smaller than isochors [emphasis added].

The potential signal in these two genomes, then, should be obvious. If not, I will belabor the point:

- The strongest correlation between mouse and rat genomes is SINE linear patterning.
- Though these SINE families have no sequence similarities, their placements are conserved.
- And they are concentrated in protein-coding genes.

Am I suggesting that extraterrestrials were fiddling with rodent DNA? No. Am I implying that we are seeing the "language of God" in rodent-script? I haven't the foggiest notion. What *I am saying* is that we know a lot about the genome that is being glossed over in the popular works that the theistic evolutionists write. I am also saying that instead of finding nothing but disorder along our chromosomes, we are finding instead a high degree of order.

Is this an anomaly? No. As I'll discuss in the next chapter, we see a similar pattern when we compare the linear positioning of human *Alus* with mouse SINEs. Is there an explanation? Yes. But to discover it, you have to think outside the BioLogos box.

15. Beginning to Decipher the SINE Signal

Richard Sternberg

REMEMBER THE ANALOGY OF THE TWO MOONS I USED EARLIER to discuss the distribution of SINEs in the mouse and rat genomes? Well, I am going to use it again today, but only for a moment.

Suppose you are keenly interested in the topography of one of the moons, named Y6-9. Suppose also that the books you first select to read on the topic are popular works, written by "experts" who are "living legends." As you read through the works, you find paragraphs here and there about how utterly decrepit Y6-9 is, and how this space body exemplifies eons of random events. The authors argue that we already knew all there was to know about that moon back in 1859, and that the evidence demonstrates either that God doesn't exist or that the deity left the cosmos to itself after the Big Bang.

You find, however, that these books almost totally ignore the findings of the billion-dollar missions sent to the surface of Y6-9 since the 1960s. Indeed, there is next to nothing in them about Y6-9's actual geology.

So you contact the LunarLogos Foundation, a Christian group that promotes such books. You tell them that you have a few specific questions about the Y6-9 mission findings. The response you get is that because you are a layman, you would not be able to comprehend the details. Besides, the LunarLogos folks say, the mainstream experts have spoken authoritatively about the subject and that should be enough for you. As a consolation, though, they send you a CD that has songs that are sung by one of their founding members.

Somewhat disgruntled, you decide to spend a day at a university library. You ask a librarian for maps of Y6-9 and technical journals that discuss its features. An hour or so later, with stacks of data before you, something catches your eye—something never mentioned in any of the books you've read. Sitting in a Y6-9 crater is a large monolith. High resolution photos reveal it to be rectangular in shape, with a polished surface, and composed of some dense black material. This must be a mistake, you think. So you look at other craters on Y6-9 and many of them also contain the same kind of monolith. You discern their overall distribution to be non-random—and the monoliths themselves are highly non-random. Then, after consulting the literature, you learn: The existence of such objects has been known for over two decades. In fact, one of the experts at LunarLogos wrote about them in the technical reports of the Y6-9 probe missions.

Now, more than disgruntled, you decide to write about what you have learned, citing the relevant literature in case someone might want to read about this topic himself. After posting what you write on the Internet, LunarLogos posts their reply. Their response reads something like this:

> Okay. Sure. There are obnoxious monoliths littering Y6-9... everybody knows this. In fact, there are about a million of them. But they got there because of degenerative cosmic processes. While many of the structures Mr. X mentioned are suggestive of some possibly unknown cause that we have never denied, it is almost certain that much, if not most, of the Y6-9 surface is without any remarkable features. Besides, why would God put them there? They are simply nonsensical.
>
> We have one more thing to say. We don't appreciate how disrespectful Mr. X has been to our team of experts. Although Mr. X is a PhD planetary scientist, he is not as qualified to write on this subject as scientists approved by LunarLogos. So we ask him, for the sake of having meaningful dialogue: Please stop writing about this subject.

A LunarLogos sympathizer writes on another blog:

> We think you're a nice guy, but your arguments are insane.

What would you think?

THEN SOMEONE UNAFFILIATED with LunarLogos brings something to your attention. He shows you a map of the sister moon of Y6-9, called Q7-10. You are aware that Y6-9 and Q7-10 went their separate ways after a cosmic collision 22 million years ago. But something strange catches your eye. Q7-10 has polished black monoliths, too, except that they are pyramids instead of rectangles. That's not the weirdest thing, though. The weirdest thing is that *the geographical distribution of the monoliths on Y6-9 very nearly matches the geographical distribution of the monoliths on Q7-10.*

Now what would you think?

The almost one-to-one correspondence of mouse-specific and rat-specific SINE insertion events along homologous regions of the two genomes is almost as remarkable as the matching geographical distributions of the monoliths in the analogy of the two moons. Remember the graph (from Figure 9c of Ref. 1):

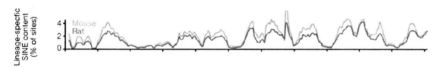

Reprinted by permission from Macmillan Publishers Ltd: figure 9c, "Genome sequence of the Brown Norway rat yields insights into mammalian evolution," *Nature* 428 (2004), pp. 493-521, copyright 2004.[33]

We have two genomes that went their separate ways 22 million years ago. We have two lineages that have been subjected to different historical events. Yet, when we compare the chromosome locations of mouse B1s/B2s/B4s with those of rat IDs, they look almost the same. Where the ID SINEs rise in density, so do the B1s/B2s/B4s SINEs; where the ID SINE levels decrease, so also do the B1s/B2s/B4s SINE levels. Independent mutational events have generated equivalent genomic patterns. How can we causally account for this striking pattern?

In the paper written by Francis Collins and his colleagues, under the heading "Co-Localization of Sines in Rat and Mouse," we read:

> The cause of the unusual distribution patterns of SINEs… *is apparently a conserved feature, independent of the primary sequence of the SINE…*" [emphasis added].

Let's unpack this part of the sentence. We have:

1. A cause of some sort.
2. A cause that is conserved between the mouse and rat.

---

33. http://www.nature.com/nature/journal/v428/n6982/full/nature02426.html

3. A cause that is independent of SINE primary DNA sequences.

That's all very well and good, but the specific cause is never mentioned. Where, then, can we find it?

Experts such as John Avise, Francisco Ayala, Francis Collins, and Darrel Falk tell us that we must think like Darwinians before we can begin to make sense of the data, since nothing else is scientific, or indeed even reasonable. So let's play along and think like Darwinians, limiting ourselves to what Collins and his colleagues have authoritatively provided. Recall that they are:

- Chance mutations continually degrade genomes that are largely junk
- SINEs are for the most part nonsensical junk
- Natural selection is the sole creative force in evolution
- Except when genetic drift (neutral evolution) is also a factor.

We can call this conceptual scheme the "BioLogos box."

We'll start, then, with chance mutations. We know that the enzymes encoded by the L1 retrotransposon copy and paste SINEs into mammalian genomes. So perhaps this is the causative agent that acts independently of primary DNA sequence? And since L1 is present in all mammalian genomes, we may just be on the trail of the "conserved cause."

But wait. L1 also mobilizes itself. This is a problem, for when we compare LINE and SINE distributions along chromosomes, it is clear that in the regions where the former is abundant the latter is not, and *vice versa*. Remember the graph (from Figure 9d of Ref. 1):

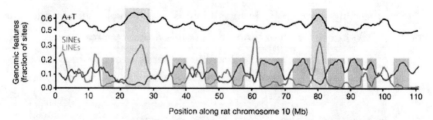

Reprinted by permission from Macmillan Publishers Ltd: figure 9d, "Genome sequence of the Brown Norway rat yields insights into mammalian evolution," *Nature* 428 (2004), pp. 493-521, copyright 2004.[34]

---

34. http://www.nature.com/nature/journal/v428/n6982/full/nature02426.html

15. Beginning to Decipher the SINE Signal

But we have no plausible mechanistic explanation for why the mouse L1 machinery would have pasted B1s/B2s/B4s—over 22 million years, no less—into the same general locations and at much the same densities, as the rat L1 machinery pasted ID elements over the same period of time.

NOT TO FEAR. We still have to consider that worker of miracles, natural selection. This mechanism eliminates harmful features while preserving those that enhance survival. So let's construct a hypothesis: Mouse and rat SINE distributions reflect the differential removal of these DNA repeats from regions where their presence would be harmful. In other words, we predict that sequences where mouse B1s/B2s/B4s and rat IDs peak in density are segments of the genome that are largely junk; conversely, in the sections where these SINEs taper off, functional coding regions are to be found.

Does this hypothesis point in the right causal direction? I don't think so. Here is why. Remember the statement made by Falk in defense of Ayala *contra* Meyer:[35]

> He [Ayala] *does* say that on average there are about 40 copies of *Alu* sequences between every two genes, but this is simply a fact.

Well, both Falk and Ayala are correct—and *that* is the problem with the selection hypothesis. Protein-coding genes make up only ~1.5 percent of the mammalian genome. Where do the peaks of B1s/B2s/B4s and IDs occur along the mouse and rat chromosomes, respectively? In and around the ~1.5 percent of the genome that is protein-coding. Remember the following statement in the sentence of the *Nature* paper quoted above:

> The cause of the unusual distribution patterns of SINEs, *accumulating in gene-rich regions* where other interspersed repeats are scarce, is apparently a conserved feature, independent of the primary sequence of the SINE... [emphasis added].

Whatever the mystery cause is, it plucked out the species-specific SINEs from the junkety-junk LINE regions, and piled them high around the "25,000 genes" of the mouse and rat. Or it directed the SINEs to rain down on the gene-rich regions and in much lesser amounts elsewhere. This contradicts our selection hypothesis, *unless the SINEs are doing something important in and around those protein-coding regions.* But since so much ink has been

---
35. http://biologos.org/blog/a-rejoinder-to-meyer-2

spilled arguing that nothing of the sort is the case—these are junk elements, even harmful—we must turn to some other factor.

Reaching into the BioLogos box, we now pull out "genetic drift." Neutral evolution means that a mutation—regardless of whether it is beneficial, neutral, or negative—can become fixed or lost in a lineage solely by chance. With respect to a SINE insertion, its persistence in a lineage would have to be a genetic coin toss: If heads, the SINE stays in a site; if tails, it is lost. So for a pure neutralist model to account for the graphs we have seen, ~300,000 random mutation events in the mouse have to match, somehow, the ~300,000 random mutation events in the rat.

What are the odds of that?

Like the imaginary scientist trying to make sense of the far-from-random lunar evidence that LunarLogos glossed over, I think we have to look elsewhere for the mystery cause of equivalent SINE patterns in the mouse and rat genomes. But where? A technical term was used in the sentence that I quoted above that you may have missed. I will highlight it for you:

> The cause of the unusual distribution patterns of SINEs, accumulating in gene-rich regions where other interspersed repeats are scarce, is apparently a conserved feature, independent of the primary sequence of the SINE and effective over regions smaller than isochores.

Ever heard of "isochores"? Well, they are to DNA sequence organization along a chromosome what mountains and valleys are to a continent. Imagine buying a book about the geographical features of Africa and not finding a single word about Mount Kilimanjaro or the Great Rift Valley.

Imagine finding instead a lot of musings about what God couldn't or wouldn't have done with Africa. What would you think about such a book?

Well, turn to the index of Francis Collins's *The Language of God*, or John Avise's *Inside the Human Genome: A Case for Non-Intelligent Design*, and look for "isochore." You won't find it.

Isochores might provide a clue to cause of the mystery signal, but the cause—whatever it may be—is outside the BioLogos box.

16. Intelligent Design, Frontloading and Theistic Evolution

Jay Richards

At Scott McKnight's blog at Beliefnet, an anonymous scientist has started a review thread[36] on Steve Meyer's book. *Signature in the Cell*. While the blogger ("RJS") says she ultimately disagrees with Meyer's argument, it's clear that she takes Meyer's argument seriously and is trying to do her best to present the argument accurately. This is much more than can be said for the many hysterical and misinformed "critiques" of Meyer's argument that are now floating around the Internet. Anyone who's actually read the book will know that most of these critiques are cliches that Meyer addresses in detail in the book, suggesting that the critics don't even know the argument they are criticizing.

A civil review like this is welcomed, and I look forward to reading the installments.

In her first installment, RJS suggests that there's a promising "third way" that Meyer doesn't address in the book:

> It seems to me that there is a middle ground between the insistence that chance, happenstance, and law (the laws of physics) suffice to explain all and the suggestion that biology—life—can only be explained with reference to a creative mind. Alister McGrath (*A Fine-Tuned Universe*) and Simon Conway Morris (*Life's Solution*) provide some insight into this middle ground. The fabric of the universe makes life possible and inevitable—not a highly contingent accident. Thinking scientifically we look for the causally connected series of

---

36. http://blog.beliefnet.com/jesuscreed/2010/01/signature-in-the-cell-1-rjs.html

events that resulted in the present reality—as part of God's method in creation.

I'm familiar with McGrath and Conway Morris's views, and they have some merit; but I don't think they offer an alternative that Meyer fails to address. Smoothing for inconsistencies in their proposals, their idea is basically that God hardwired or "frontloaded" everything "in the beginning" as it were to give rise to complex life somewhere, while allowing for a lot of "freedom" and variation within the cosmos. (So they're not hard determinists.)

First, taken seriously, this is quite obviously a theistic form of design that simply tries to locate all the designing activity at the beginning—in the cosmic fine-tuning and initial conditions. The design does real work, and there's no reason that the effects of that design would not be empirically detectable (as long as we have an open-minded, nonpositivist view of science). As a simple analogy, think of frontloading this way. If I shoot a gun at a target and hit it, I've intentionally aimed the bullet at the beginning, even though the bullet's trajectory follows the rules of gravity, momentum, etc. In God's case, of course, he would also establish the law-like rules and superintend them. All I can do when I shoot a gun is take them into account.

Second, some frontloading and fine-tuning is not only compatible with but necessary for Steve's argument. But I think the argument that everything can be explained this way doesn't capture the details of Steve's argument about information at the origin of life. The frontloading scenario tries to turn necessary conditions for life into sufficient conditions. Though Steve doesn't say this, if he's right, it's not at all obvious that this frontloading scenario is so much as possible. The only thing God would have to hardwire information at the beginning would be initial conditions, some proto-matter and the repetitive, law-like forces that govern the matter. But we can see the effects of both those initial conditions and the law-like regularities playing out in the material world now. They constitute the *background* to the information in biological systems—that is, the necessary but nowhere nearly sufficient background—the contrast medium for the information. What would it mean to tweak the expressions of gravity and electromagnetism so that they would give rise to the information-processing in cells and body plans of vertebrates? I think this explanation has plausibility only in proportion to the haziness of one's conception of specified information.

Third, even if it's possible for God to frontload things in this way, it hardly follows that this is a better explanation than the one Steve proposes, which is (at least implicitly) (1) that matter shows degrees of freedom inconsistent with such complete frontloading and (2) that intelligence plays an active and detectable role within cosmic history, and probably is not limited in the way proposed (or suggested) by Conway Morris and others. What we're interested in is the best explanation for life's features in the real world, one that takes account of the known causal powers of the world as we see it.

17. Getting ID Right: Further Thoughts on the Beliefnet Review

Jay Richards

The SECOND, THIRD, AND FOURTH INSTALLMENTS OF THE REVIEW of Steve Meyer's book *Signature in the Cell* are up over at Beliefnet.[37] (I responded to the first installment in the preceding chapter.)

Although this series appears on Scot McKnight's Jesus Creed blog, they're written by anonymous blogger "RJS." I'm guessing that RJS is a scientist, or is in a sensitive academic position, and doesn't want to risk banishment for saying reasonable things about an ID argument. If so, that tells us something of the social pressures against writing publicly about this issue.

The second installment didn't really review Meyer's book, but rather used Meyer's analysis of evidence in the historical sciences as a point of departure for reflecting on the differences in historicity between Noah's flood and Jesus's resurrection. She has some interesting thoughts on this, but since it's not germane to Meyer's argument, I'll just respond to her third installment here, and her fourth and later installments separately.

I should say that this review is better than 95 percent of online reviews of Meyer's book, so it's worth reading. Not only does she grapple with the details, she's actually read the book before reviewing it. What a thought!

Unfortunately, she still mischaracterizes ID, and she still relies on the Darwinian doctrinal defaults so characteristic of this debate.

First, she makes it appear that ID is concerned only with the biological sciences, which is not the case. It's just that biology is by far the most contro-

---

37. RJS's three later review installments may be accessed here: http://blog.beliefnet.com/jesuscreed/2010/01/signature-in-the-cell-2---hist.html; here: http://blog.beliefnet.com/jesuscreed/2010/01/signature-in-the-cell-3---hist.html; and here: http://blog.beliefnet.com/jesuscreed/2010/01/signature-in-the-cell-4-rjs.html.

versial area for saying design-friendly things (due to the deeply ideological character of modern neo-Darwinism), so it draws the most fire.

Second, though I'm glad she distinguishes the negative case against, say, neo-Darwinism, from the positive case for intelligent design, she puts the point a bit pejoratively as "the attempt to undermine all of evolutionary biology." When dealing with the negative side of the argument, the focus among ID folks in biology is not "all of evolutionary biology," but rather the Darwinian selection-mutation mechanism, materialistic chemical origin-of-life scenarios, and inaccurate claims concerning universal common ancestry. And IDers widely recognize that it's the first two claims, and not the third, that are central to the argument.

But there is an issue in this vicinity that RJS misses: if you're allowed to consider ID, then many arguments for (universal) common ancestry are ambiguous, and seem to count equally in favor of common design and common descent. ID folks generally understand this and are willing to talk about it publicly, while those seized with the Darwinian vision usually find it almost impossible to imagine the evidence for common descent counting for anything else. RJS does this almost reflexively, citing just this sort of ambiguous evidence from Darrell Falk and Francis Collins for this conclusion:

> These three lines of evidence, and perhaps there are others, make the general theory of evolution clearly the inference to best explanation. There is no real doubt left. While we do not yet understand the whole process, the general scenario is as close to proven as anything ever is or can be in history or biology. Arguments against the broad brush history of evolution fall into the same general category as arguments that Napoleon never existed (an example Meyer uses in his book when discussing IBE), that Jesus was married, or that the Holocaust never happened.

These sorts of doctrinal statements are nearly universal in this debate, and should always set off your baloney detector. This one doesn't even pass the smell test. "The broad brush history of evolution" is hopelessly ambiguous. Are we talking about history, change over time, cosmic evolution, universal common ancestry, or all of the above plus the mutation-selection mechanism and other putative mechanisms that are often referred to vaguely but seldom do any real work in creating adaptive complexity? We're not told. And whatever it encompasses, it surely involves all sorts of different claims and inferences about deep history. As a result, even if it were precisely defined, it would still be qualitatively different from discrete events in very recent, recorded human history. Alas, such comparing of apples and orangutans is common in the evolution debate, and serves no helpful function.

THE REVIEWER'S CONCERN here does seem to be of the "helpful advice" variety: ID would have more credibility if it would drop all the snake-handling stuff: "I think that the ID movement damages its credibility (destroys might be a better word) by fighting a battle against the general evolutionary theory." But that's the reviewer's misleading characterization of ID, based in part on her apparent confusion about the differing status of different historical events. I am surprised that RJS makes this mistake, since her second installment was an excursus on the intrinsic differences in the historicity of Noah's flood and Jesus' resurrection. Thus she can draw careful distinctions. And yet, when we move to evolutionary theory, this capacity for nuance reverts to default invocations about the impeccable evidential credentials of some ill-defined evolutionary scenario.

What this suggests to me is that there's something about the logical and rhetorical character of the "general evolutionary scenario" that makes it very hard for those enamored of it to keep separate issues separate.

III

ATTACK OF THE PYGMIES

18. Signs of Desperation? Early Responses to *Signature in the Cell*

Casey Luskin

IF THE STRENGTH OF AN ARGUMENT IS REFLECTED IN THE FEEBLENESS of the rebuttals it receives, then Stephen Meyer's manifesto, *Signature in the Cell: DNA and the Evidence for Intelligent Design*, might be a rare rhetorical gem.

In six hundred pages, Meyer takes apart many of the leading materialistic theories for the origin and evolution of life with an unrelenting barrage of logic and evidence, yet also with respect for his opponents. As Heather Zeiger aptly commented in the journal *Salvo*, "The value of his book is not merely in its conclusion that intelligence best explains the source of the DNA code; it is in the process Meyer uses to bring us to this conclusion. The reader sees the scientific process firsthand."[38]

But have Meyer's critics responded with such grace and rhetorical punch?

After debating Stephen Meyer on the Michael Medved radio program last November, science journalist Chris Mooney apparently felt he couldn't find sufficient ammo to rebut the Cambridge-trained philosopher of science. Thus, Mooney subsequently posted a piece on his *Discover Magazine* blog, titled "Time to Refute Stephen Meyer?", in which he lamented that "Meyer's book is clearly drawing a lot of attention and is scarcely being refuted so far as I can see."[39]

Mooney was correct that Meyer's book was garnered much interest—though not from critics. In November 2009, an endorsement from the leading political philosopher (and atheist) Thomas Nagel led to its being named

38. http://www.salvomag.com/new/articles/salvo11/11zeiger.php
39. http://blogs.discovermagazine.com/intersection/2009/11/17/time-to-refute-stephen-meyer/

one of the "Books of the Year" by the prestigious *Times Literary Supplement* in London. The following month, Meyer was named "Daniel of the Year" by *World Magazine* for the "courage" and "perseverance" that led to *Signature in the Cell*.

Around this time, the anti-ID community on Internet decided they could not afford to continue ignoring Meyer's book, and critical reviews began trickling in. In the spirit of respectful scholarly debate, I will assess some of the counter-arguments and give five friendly tips to critics of Stephen Meyer.

First, know the *man you're attacking*. University of Chicago evolutionary biologist Jerry Coyne tried to dismiss Meyer as a young-earth creationist and had to retract his claim.[40] Had Coyne read Meyer's book, he would have learned that Meyer's views about the age of the earth were no secret. Not only does *Signature in the Cell* adopt the mainstream geological time scale, but as long ago as the 2005 Kansas science hearings, Meyer plainly stated, "I think the earth is 4.6 billion years old. . . . That's both my personal and my professional opinion."

A second tip for critics of *Signature in the Cell* is to *read the book before reviewing it*. In December 2009, biology professor P.Z. Myers directed readers of his heavily trafficked blog to a call for negative reviews of *Signature*—while simultaneously declaring, "I suppose I'll have to read that 600-page pile of slop sometime . . . maybe in January."

Seeing that their leader was publicly attacking a book he hadn't read, P.Z.'s fans felt justified in doing the same. Amazon.com saw a sudden spike in short, negative one-star reviews of *Signature in the Cell* that had little to do with any of the arguments in the book.

The smear campaign, however, did not have its intended effect. By the end of 2009, *Signature in the Cell* was fast becoming one of the bestselling science books of the year on Amazon.

---

40. http://whyevolutionistrue.wordpress.com/2009/07/15/pro-intelligent-design-editorial-in-boston-globe/

A third mistake—particularly common among critics who didn't heed my second tip—is to cast Meyer's argument for design as a mere negative critique of evolution. Instead, *try to stay positive.*

For example, P. Z. Myers caricatured the book by stating, "I know what is in this book—'ooooh, it's so complex, it must have been . . . DESIGNED!'" Had P. Z. read the book, he would have discovered a rigorous positive case for design based upon finding in nature the precise type of information that, in our experience, comes from intelligence.

"What humans recognize as information certainly originates from thought—from conscious or intelligent human activity," writes Meyer in the opening chapter of his book. "Our experience of the world shows that what we recognize as information invariably reflects the prior activity of conscious and intelligent persons." Later in *Signature*, Meyer elaborates on the precise type of information that reliably indicates the prior action of an intelligent cause:

> Experience shows that large amounts of specified complexity or information (especially codes and languages) *invariably* originate from an intelligent source—from a mind or personal agent. . . . So the discovery of the specified digital information in the DNA molecule provides strong grounds for inferring that intelligence played a role in the origin of DNA. Indeed, whenever we find specified information and we know the causal story of how that information arose, we always find that it arose from an intelligent source.

Chris Mooney must also have skipped over Meyer's carefully laid out, positive argument for design (which is hard to miss, since it is woven through the entire book). Mooney claims that Meyer merely "throws up his hands, and says, it's so improbable, God must have done it."

With this gross misrepresentation of Meyer's argument, Mooney follows the same approach he took in *The Republican War on Science*, in which he claimed that a peer-reviewed scientific paper authored by Meyer was "lacking" a "positive case for the necessity of ID." But in that paper, Meyer had argued that "design theorists are not positing an arbitrary explanatory element unmotivated by a consideration of the evidence" but instead are "positing an entity possessing precisely the attributes and causal powers that the phenomenon in question requires as a condition of its production and explanation."

FEW CRITICS OF ID seem capable of following my fourth tip: *Remain civil*—or at least make some minimal attempt at civility. Unsurprisingly, P. Z. Myers leads the charge in violating this ground rule of discourse, calling *Signature in the Cell* "Discovery Institute Bulldung," and proclaiming that "Stephen Meyer lies."[41]

While readers of P. Z.'s blog generally cheer on his every invective, readers of Jerry Coyne's blog respond a little differently. After Coyne called Meyer a "Discovery Institute creationist and lying liar," one of Coyne's readers commented, "Meyer seems like a lot of things—including smart—but I don't think he is a deliberate liar. He appears to be a nice guy who differs with you about some things. Attributing malign motives to others only serves to demonize them and make dialogue more difficult." Coyne did not reply.

While these anecdotes are revealing, an informal survey by Tom Gilson, who runs the popular blog ThinkingChristian.net, actually tried to quantify the level of civility and open-mindedness among various reviewers. His findings were striking. Among the negative, one-star reviewers of *Signature in the Cell*, Gilson found that "more than nine-tenths said something to the effect that the question is settled, there's no need to pursue it anymore. Many of them were more colorful than that: *The question is settled, and attempts to keep pursuing it are just lies from the 'Dishonesty Institute.'*"

But Gilson found that "those who rated the book highly had more open minds to the issue: only 20 percent of that group made statements to the effect that 'the question is now settled.'" This seems to counter James Madison University mathematics professor and "new atheist" Jason Rosenhouse, who asserted when reviewing *Signature in the Cell*, "Phony claims of certainty are far more typical of religion than they are of atheism."

A fifth common mistake made by critics of *Signature in the Cell* is to attempt theological rather than scientific rebuttals. Much better to *stick to the science*. Francisco Ayala, an eminent evolutionary biologist and former Catholic priest, who once served as president of the American Association

---

41. http://scienceblogs.com/pharyngula/2009/07/more_discovery_institute_bulld.php

for the Advancement of Science, critiqued *Signature in the Cell* on the website of the BioLogos Foundation.

Ayala proclaimed that ID is tantamount to "blasphemy" because it implies that God is responsible for "design defects," such as tsunamis, back aches, misaligned teeth, and complications encountered during childbirth. Ayala's argument for Darwinism is almost entirely theological: "people of faith would do better to attribute the mishaps caused by defective genomes to the vagaries of natural selection and other processes of biological evolution, rather than to God's design."[42]

One could flippantly note that orthodontists and chiropractors may in fact rejoice over such "design defects," but a serious response to Ayala can be made just as succinctly. Jay Richards did so in *Salvo* ("Can ID Explain the Origin of Evil?"). "'Bad designs' and 'evil designs' are still designs; neither of these arguments refutes ID," he pointed out. "The problem of evil isn't an argument against ID. An argument for intelligent design is just that. Questions about evil and about the nature of the designer are separate questions." Meyer corroborates this point in *Signature in the Cell*, writing, "Though the designing agent responsible for life may well have been an omnipotent deity, the theory of intelligent design does not claim to be able to determine that."

AYALA'S READERS AT BioLogos wasted little time in spotting these fallacies. "Dr. Ayala appears to be one of the many reviewers who have not read Dr. Meyer's book," wrote the first commentator on the review. "If he has read it, he has not explained why he chose not to address any of the main arguments Meyer makes in the book." The reader went on to say that Ayala does "not seem to understand Intelligent Design" because he goes "on and on about 'bad design' in nature, without showing any awareness of the responses to such arguments that design proponents have made for many years." The commenter concluded, "This does not further the debate."

The BioLogos Foundation itself is a theistic evolution advocacy group founded by Francis Collins, which prompts the question: Should religious persons trust the *theology* of ID critics like Ayala on topics like God, natural evil, and design? In a 2008 interview, the *New York Times* reported that Dr.

---

42. http://biologos.org/blog/on-reading-the-cells-signature/

Ayala wouldn't say whether or not he remained a religious believer, because, in Ayala's words, "I don't want to be tagged... by one side or the other." Thus, Ayala represents perhaps the most eminent proponent of the view that ID is bad theology—and apparently is endorsed by the theistic evolutionists at BioLogos as a spokesman in the debate—yet he categorically refuses to say publicly whether he is a religious believer or not.

Not only that, but he missed the mark by miles when responding to Meyer.

The public rebuttals of *Signature in the Cell* may be inadequate, but does this mean that materialists will never explain the origin of information in the cell? Not at all. In scientific debates, one must always remain open to future discoveries. But the showing thus far does mean that intelligent design deserves serious scientific consideration—not abrasive quips, dismissals, and refusals to engage Meyer's arguments.

Undoubtedly, more reviews of Meyer's book are forthcoming. Nonetheless, as one reviewer on Amazon put it well: "If materialists continue to fail to answer Meyer's arguments, or even to seriously engage them, then the tipping point in the debate over design in biology is close at hand."

19. Get Smart: Can Unintelligent Causes Produce Biological Information?

Casey Luskin

In the preceding chapter, the final tip I gave to reviewers of Stephen Meyer's *Signature in the Cell* was "stick to the science." While many reviewers unashamedly boasted of not having read the book (or wrote rebuttals so far askew from Meyer's argument that one could not help but question whether they had), a few critics have published serious scientific responses. This chapter will assess some critics who—though not always refraining from personal attacks and misrepresentations—at least attempted scientific rebuttals to Meyer sufficient to warrant response.

"Two things struck me as I read [*Signature*]," wrote University of Waterloo mathematician Jeffrey Shallit in response to Meyer in January 2010. "First, its essential dishonesty, and second, Meyer's significant misunderstandings of information theory."[43]

Setting aside the gratuitous invective, Shallit's main objection is that Meyer defines information as "specified complexity," rather than Shannon information or Kolmogorov complexity, the terms which he elsewhere claims are the "accepted senses of the word as used by information theorists." Shallit is so opposed to the ideas advocated by ID proponents that he even refuses to use Meyer's term, "specified complexity," instead calling it "creationist information."

---

43. http://recursed.blogspot.com/2009/10/stephen-meyers-bogus-information-theory.html

But in *Signature in the Cell*, Meyer spends pages explaining why Shannon information in fact is not a useful measure of functional biological information. Meyer first asks us to consider two sequences of characters:

String A: "Four score and seven years ago"

String B: "nenen ytawoi jll sn mekhdx nnx"

Since "both of these sequences have an equal number of characters," explains Meyer, "both sequences have an equal amount of information as measured by Shannon's theory." The logic is simple but devastating: "one of these sequences communicates something while the other does not," and therefore "Shannon's theory cannot distinguish functional or message-bearing sequences from random or useless ones."

When measuring the related concept of Komolgorov information, the problem is even worse. Komolgorov complexity can be thought of this way: What is the minimum length of a computer program needed to generate a string? The more commands necessary, the greater the Kolmogorov complexity. Yet under Kolmogorov complexity, a stretch of completely functionless junk DNA that has been utterly garbled by random, neutral mutations might have *more* Kolmogorov complexity than a functional gene of the same sequence length.

To understand why, consider the two sequences above. Since many of the characters in the first string could be predicted using the grammatical rules of English, it actually has *less* Kolmogorov complexity than String B, which is randomized. Yet clearly String A conveys far more meaningful information than String B.

For obvious reasons, neither Shannon nor Kolmogorov information are useful metrics of functional biological information; a useful measure of biological information must take into account the function specified by the information. And despite Shallit's vitriolic assertions to the contrary, he seems unaware that Meyer's use of the term "specified complexity" (also called "complex and specified information," or CSI) is supported by eminent scientists who are by no means "creationists."

I N A 1973 book cited by Meyer, *The Origins of Life: Molecules and Natural Selection*, leading origin-of-life theorist Leslie Orgel—a staunch materialist—described "specified complexity" as a hallmark of the information in living organisms:

> [L]iving organisms are distinguished by their specified complexity. Crystals are usually taken as the prototypes of simple, well-specified structures, because they consist of a very large number of identical molecules packed together in a uniform way. Lumps of granite or random mixtures of polymers are examples of structures which are complex but not specified. The crystals fail to qualify as living because they lack complexity; the mixtures of polymers fail to qualify because they lack specificity.[44]

When responding to Meyer's recommendation that we measure biological information in terms of the specification necessary to perform some function, Shallit asserts, "This is pure gibberish. Information scientists do not speak about 'specified information' or 'functional information.'"

Again, Shallit must be unaware that leading scientists have used those very terms while simultaneously arguing that classical information theory is not useful for measuring biological information.

In 2003, Nobel Prize-winning origin-of-life researcher Jack Szostak wrote a review article in *Nature* lamenting that the problem with "classical information theory" is that it "does not consider the meaning of a message" and instead defines information "as simply that required to specify, store or transmit the string."[45] According to Szostak, "a new measure of information—functional information—is required" in order to take account of the ability of a given protein sequence to perform a given function.

Some theorists are heeding Szostak's call for better definitions of functional biological information. A 2007 paper in the journal *Theoretical Biology and Medical Modelling* found that some measures of biological complexity are not "sufficient to describe the functional complexity observed in living organisms" and instead recommended measuring biological information through functional sequence complexity (FSC):

> FSC includes the dimension of functionality. Szostak argued that neither Shannon's original measure of uncertainty nor the mea-

---

44. Orgel, Leslie E., *The Origins of Life: Molecules and Natural Selection* (Chapman & Hall, 1973), p. 189.
45. Szostak, Jack W., "Molecular messages," *Nature* 423:689 (2003).

sure of algorithmic complexity are sufficient. Shannon's classical information theory does not consider the meaning, or function, of a message. Algorithmic complexity fails to account for the observation that "different molecular structures may be functionally equivalent." For this reason, Szostak suggested that a new measure of information—functional information—is required.[46]

In 2007 Szostak co-published a paper in *Proceedings of the National Academy of Sciences*, with Carnegie Institution origin-of-life theorist Robert Hazen and other scientists, furthering these arguments. Attacking those like Shallit who insist on measuring biological complexity using the outmoded tools of classical information theory, the authors wrote, "A complexity metric is of little utility unless its conceptual framework and predictive power result in a deeper understanding of the behavior of complex systems." Thus they "propose to measure the complexity of a system in terms of functional information, the information required to encode a specific function."[47]

Clearly Meyer's arguments have a strong precedent in the scientific literature. Shannon information and other metrics used by classical information theory are insufficient for use by many biologists because they fail to take into account the complexity needed to produce a biological function. Whether the metric is called "functional information," FSC, or CSI, scientists are finding better ways of measuring biological complexity than Shannon information.

Some critics, however, were far more civil than Shallit. Point Loma Nazarene University biology professor Darrell Falk accused Meyer of hindering the progress of science by reaching "premature conclusions based on his unsuccessful attempt to move from philosophy into genetics, biochemistry and molecular biology."

Falk cited two papers to justify his argument. The first was published just before the release of *Signature in the Cell*, which purportedly showed, ac-

---

46. Durston, Kirk K., David K. Y. Chiu, David L. Abel and Jack T. Trevors, "Measuring the functional sequence complexity of proteins," *Theoretical Biology and Medical Modeling* 4:47 (2007) (internal citations removed).
47. Hazen, Robert M., Patrick L. Griffin, James M. Carothers and Jack W. Szostak, "Functional information and the emergence of biocomplexity," *Proceedings of the National Academy of Sciences, USA*, 104 (2007), pp. 8574–8581.

cording to Falk, "a very feasible way" that two RNA nucleobases "could have been produced through natural processes."

However, some eminent origin of life theorists disagree. When commenting on this research last May, Robert Shapiro, professor emeritus of chemistry at New York University, stated, "The chances that blind, undirected, inanimate chemistry would go out of its way in multiple steps and use of reagents in just the right sequence to form RNA is highly unlikely." The research, said Shapiro, "definitely does not meet my criteria for a plausible pathway to the RNA world" because one of the "assumed starting materials is quickly destroyed by other chemicals and its appearance in pure form on the early earth 'could be considered a fantasy.'"[48]

Commenting for *Nature*, Shapiro further argued, "The flaw is in the logic—that this experimental control by researchers in a modern laboratory could have been available on the early Earth." Though Shapiro wouldn't put it this way, the problem is that producing these nucleobases requires intelligent design. As a pro-ID chemist commented to me privately about this research, "The work was very carefully done. The problem is that it was very carefully done."

Falk also argued that Meyer's position was refuted in a paper by Gerald Joyce, who produced a self-replicating RNA molecule in the laboratory. Apparently Falk was unaware that Meyer publicly responded to both of these research papers before his book was released.

"The central problem facing [origin of life research] is not the synthesis of pre-biotic building blocks or even discovering an environment in which life might have plausibly arisen—difficult as these problems have proven to be," wrote Meyer in June of 2009 on Discovery Institute's blog, Evolution News & Views. "Instead, the fundamental problem is getting the chemical building blocks to arrange themselves into the large information-bearing molecules (such as DNA and RNA) that direct the show in living cells."

---

48. Wade, Nicholas, "Chemist Shows How RNA Can Be the Starting Point for Life," *New York Times* (May 14, 2009).

Meyer's rebuttal charged that this "information sequence" problem is unsolved by Joyce's replicating RNA molecules, which required intelligent engineering to properly order their information content:

> [S]ignificantly, Joyce intelligently arranged the matching base sequences in these RNA chains. Thus as my forthcoming book, *Signature in the Cell*, shows, Joyce's experiments not only demonstrate that self-replication itself depends upon information-rich molecules, but they also confirm that intelligent design is the only known means by which information arises.

The fundamental thesis of Meyer's paper—that the ordered information in DNA requires an intelligent cause—remains untouched by Joyce's innovative, though ultimately inadequate research. Unless materialists can demonstrate that unguided, unintelligent material causes can generate new complex and specified information in biology, Meyer's thesis is going to stand strong for a long time.

20. Weather Forecasting and Complex Specified Information

Paul Nelson

For over a decade, mathematician Jeffrey Shallit has been an outspoken critic of intelligent design. Recently, in a series of blog posts, he has attacked Stephen Meyer's book *Signature in the Cell* for what he sees as a variety of shortcomings.[49] Some of Shallit's criticisms merit careful attention.

Other criticisms, however, are fluffy confections, failing to achieve even the slightness of what Hume called "mere cavils and sophisms." Let's look at one such bonbon of sophistry: Shallit's claim that weather forecasting represents a devastating counterexample to *SITC*'s argument that complex specified information is, universally in human experience, produced by a mind or intelligence.

Shallit writes:

> Even if we accept Meyer's informal definition of information with all its flaws, his claims about information are simply wrong. For example, he repeats the following bogus claim over and over:
>
> p. 16: "What humans recognize as information certainly originates from thought—from conscious or intelligent human activity... Our experience of the world shows that what we recognize as information invariably reflects the prior activity of conscious and intelligent persons." [...]
>
> I have a simple counterexample to all these claims: weather prediction. Meteorologists collect huge amounts of data from the natural world: temperature, pressure, wind speed, wind direction, etc., and

---

49. Shallit's blog posts may be accessed here: http://recursed.blogspot.com/2009/10/stephen-meyers-bogus-information-theory.html; and here: http://recursed.blogspot.com/2010/01/more-on-signature-in-cell.html.

process this data to produce accurate weather forecasts. So the information they collect is "specified" (in that it tells us whether to bring an umbrella in the morning), and clearly hundreds, if not thousands, of these bits of information are needed to make an accurate prediction. But these bits of information do not come from a mind—unless Meyer wants to claim that some intelligent being (let's say Zeus) is controlling the weather. Perhaps intelligent design creationism is just Greek polytheism in disguise!

Poor Zeus: stand-in deity for yet another counterexample. And he only gets union scale.

To see what's wrong with this putative counterexample, begin by asking yourself if you know—without peeking online at a weather page, or looking at a barometer—what the atmospheric (barometric) pressure happens to be in your immediate neighborhood, right now.

Any guesses? Well, how about the temperature, or the wind speed and direction? Again, no peeking allowed. Give yourself a moment or two to write down the correct values. Okay, stop.

The fact is, unless you cheated, you don't know the relevant measurements for your immediate surroundings (nor do I, as I write this, for my neighborhood). So what would we need to obtain those data?

Measuring instruments, such as (a) a barometer, (b) a thermometer, (c) a wind speed indicator (an anemometer), (d) a wind direction indicator, and so on. So let's suppose we have these instruments, and we retrieve data from all of them.

Can we now predict tomorrow's weather? Do we have, as Shallit argues, complex specified information?

No. We have a few data points. To create an accurate weather prediction, we're going to need data retrieved from hundreds or thousands of instruments, distributed or coordinated across a wide geographic area, and taken over a range of time intervals.

We're going to need something more, however, without which all those measurements will tell us nothing. We need an analytical model—an algorithm—and a computer to run that model.

20. Weather Forecasting and Complex Specified Information

Shallit glides over this essential step in how data become predictions with his innocent, almost blushingly naïve verb, "process":

> Meteorologists collect huge amounts of data from the natural world: temperature, pressure, wind speed, wind direction, etc., and *process* this data to produce accurate weather forecasts [emphasis added].

Now, "process" can mean many things. What "process" manifestly does not mean in the case of weather forecasting, however, is the mechanical transmission of thousands of bits of data directly from measuring instruments to end-users. That would look like this:

```
WASHINGTON REGIONAL WEATHER ROUNDUP
NATIONAL WEATHER SERVICE SEATTLE WA
700 PM PDT SUN MAY 09 2010
SEATTLE, TACOMA, EVERETT, AND VICINITY

CITY SKY/WX TMP DP RH WIND PRES REMARKS
EVERETT MOSUNNY 58 43 57 VRB7 29.85F
BOEING FIELD MOSUNNY 69 17 13 NE3 29.82F
RENTON MOSUNNY 66 27 23 CALM 29.82F
SEATAC AIRPORT MOSUNNY 66 25 21 W3 29.83F
MCCHORD AFB MOSUNNY 65 37 35 W12G22 29.83F
PUYALLUP FAIR 66 32 28 CALM 29.86F
BREMERTON MOSUNNY 63 30 29 SW8 29.86F
```

Or this:

```
                         Wind
Time RHum      Temp Direct Speed Gust Rain Radiation Pres
         (%) (F) (knot) (knot) (inch) (Watts m-2) (mbar)

00:00:29 16 66 213 5 6 0.00 479.17 1010.9
00:01:29 16 66 210 5 5 0.00 475.03 1010.9
00:02:29 16 66 234 4 4 0.00 473.25 1010.9
00:03:29 16 66 260 4 5 0.00 472.38 1010.9
```

And we'd be none the wiser. There is a reason we don't receive our weather predictions this way: *raw data aren't predictions at all.* Collecting measurements from instruments, and mechanically transmitting those data, without any interpretation or analysis, does not (indeed, cannot) make any specified predictions.

To be sure, there is complexity aplenty in the data, but, as *SITC* explains, that complexity is unspecified. Unspecified complexity is what natural causes produce. And thus, because that "information" lacks specification, it is useless (by itself) for yielding genuine predictions. No specificity; no forecast.

By contrast, in real weather forecasting, data only become complex specified information—that is, genuine predictions—by passing through an intelligently designed algorithm: a computer model, in most instances. But the measurements themselves don't give us the model. They can't.

Metrologists construct models, using their minds (their analytical intelligence). The useful, complex specified information that emerges from a computer model comes to us via the action of intelligent agents, and not otherwise. The true "process" therefore looks like this:

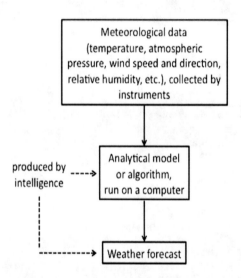

S HALLIT WOULD SUCCEED if he could show how raw meteorological data yield testable weather forecasts, without those data ever passing through the analytical filter of an intelligently designed model or algorithm.

Good luck with that.

One final point. In other writings, Shallit has indicated his hostility to the notion of human agency.[50] In light of this, it's perhaps not surprising that Shallit reduces the creative intellectual activity of meteorologists, who can

---

50. http://recursed.blogspot.com/2010/01/free-will-being-challenged.html

improve their predictions by designing better and more powerful algorithms, to the bland and seemingly agency-free verb, "process."

But unpacking that verb shows clearly that intelligent causation is actually indispensable, whether Shallit sees it or not.

Support your local meteorologist.

21. Gotcha! Stephen Meyer's Spelling & Other Weighty Criticisms

Casey Luskin

What would you get if you crossed a snarky pro-evolution blog like Panda's Thumb with a passionate defender of theistic evolution? You might get the critique of Stephen Meyer's book *Signature in the Cell* (SITC) written by biology professor Steve Matheson of Calvin College. On his personal blog,[51] Matheson has been reviewing *SITC* chapter by chapter, mixing frequent personal attacks on Meyer with exposés of occasional typos and the possible discovery of one minor error. That is not a bad track record for Meyer, considering that at the writing of this response, Matheson has reviewed nearly half the book.

Matheson is noteworthy because he at least gives every indication that he's reading *Signature in the Cell* before attacking its author. It would have been preferable for Matheson to have read the book entirely before rendering judgment. But when it comes to many other critics of *Signature in the Cell* on the internet, *this is progress*.

Unfortunately, Matheson feels it necessary not just to critique *SITC* but to smear it as "not a serious work of scholarship," not "serious science," "awfully bloated," potentially "a joke," "disingenuous," "sad," "pathetic," and "fluffy and vacuous, simplistic at best and not infrequently wrong or misleading." In case you didn't get the point, Matheson accuses Meyer of "some combination of ignorance, sloth, and duplicity," using tactics that require "layers of dishonesty" that is "sufficient to justify a charge of deliberate dishonesty."

---

51. All posts taken from Steve Matheon's blog, "Quintessence of Dust," at http://sfmatheson.blogspot.com/.

So what, in his eyes, is so bad about the book?

Of Meyer's Chapter 1, Matheson opens by saying the title ("DNA, Darwin, and the Appearance of Design") is "a poor start," followed by a "sloppy" and "fluffy chapter." Matheson tries to dismiss Meyer's argument as one of "incredulity," writing: "To establish the kind of design claim that Meyer wants to make, one must do more—much, much more—than merely pointing to current scientific ignorance or (worse) confessing personal incredulity in the face of proposed scientific explanation."

This is a rhetorical trick: *SITC* spends pages giving scientific and logical rebuttal to various theories for the origin of life and life's information, but Matheson re-labels and dismisses this lengthy argument as merely showing Meyer's "incredulity." *Perhaps incredulity is warranted if years of investigation show a paradigm is incapable of explaining the data. Since when is it a scientific sin to point out that a particular class of models does not account for our observations?*

Matheson contends that "what one must do is show that the non-design alternative (whatever it is) is unable to provide the expected explanation." He ignores the fact that this is a major part of what Meyer does throughout the book.

If you're beginning to suspect that Matheson is determined to find fault with Meyer's book regardless of the facts, then you're on to something. Matheson goes on to accuse Meyer of making an "embellishment" that is "pathetic." What's Meyer's offense? Meyer said that The Philosophical Library (which published the groundbreaking ID book *The Mystery of Life's Origin*) is "a prominent New York publisher of scientific monographs." Matheson claims that visiting the website of The Philosophical Library refuted Meyer's claim, but a glance at the website in fact shows they've published books from Nobel Prize winners in Chemistry, Physiology or Medicine, and Physics, including various titles by Albert Einstein.[52] Meyer's claim doesn't seem like an "embellishment" at all.

But we need to step back for a moment: Why are we even talking about this? It's part of Matheson rhetorical strategy, cribbed from the Panda's-

---
52. See http://www.philosophicallibrary.com/.

Thumb playbook: He wants readers to feel that Meyer can't be trusted, so he tries to smear him. Yet it's Matheson's charges and accusations that disintegrate on even the most cursory inspection.

Matheson begins his review of Chapter 2 by *praising* Meyer but soon thereafter says he is "disgusted" by what he calls "a return to rhetorical tactics" of Chapter 1. Why the sudden fury? It's because Meyer says the following: "The problem of the origin of life had at last been solved. Or at least so it seemed, until scientists began to reflect more deeply on the other great discovery of 1953." Matheson explains his disgust:

> That last sentence suggests that "scientists" share Meyer's seeming awe before "the DNA enigma." And I don't believe that at all.

Perhaps Matheson wishes to believe that no "scientists" doubt the ability of material causes to produce the information in DNA. But that doesn't mean that these scientists—many of whom are cited throughout *SITC*—don't exist.

Throughout his blog posts on *SITC*, Dr. Matheson shows what a good copy editor he would make. (For other people's writing, anyway; his own blog isn't devoid of typos.) Thus on page 66 of *SITC*, there's one instance where Meyer writes "virus" when he should have written "bacteria." Matheson is ready to pounce. Everywhere else in the passage (including over ten instances) Meyer *correctly* writes "bacteria," so it's pretty obvious this is just a typo. Normally reviewers would ignore trivial mistakes like this, especially since, once again, Meyer correctly wrote "bacteria" everywhere else in the passage. But Matheson jumps at the opportunity to underline Meyer's supposed incompetence, proclaiming:

> But this biologist finds the error more significant, and I suspect others would agree. The difference, I think, is that I can't imagine mistaking a virus for a bacterium; it's like mistaking a pencil for a sequoia.

Matheson must really be desperate to find fault with *SITC*. There's no indication that Meyer doesn't know the difference between a bacterium and a virus. It's a typo that even a biologist could make (unless Matheson believes biologists are inerrant when typing about the subject of biology).

In Chapters 4 and 5, Matheson thinks he finds multiple mistakes which he calls "embarrassing," "devastating," even implying "dishonesty." But close analysis shows he only finds one potential error —a minor one that in no way affects Meyer's overall argument.

Matheson faults Meyer's description of an experiment where Meyer wrote that mice died due to "proteins that were toxic" (p. 104), when in reality DNA transferred into the bacteria produced an enzyme that gave the bacteria a polysaccharide coating. This didn't itself kill the mice, but rather prevented the mouse immune system from being able to detect and destroy the bacteria.

While Meyer could have made that point clearer, it *was* ultimately toxic proteins from the bacteria that killed the mice—most likely pneumolysin.[53] So it's not clear that Meyer was actually in error, although he could have clarified that the transferred DNA didn't produce the toxic protein but rather merely enabled bacteria to evade the mouse immune system. But the mice still died, just as Meyer says, due to "proteins that were toxic."

Matheson goes on to obsess about another typo: Meyer's misspelling of "spliceosome" as "splicesome" on one page of his book (the term is spelled correctly on two other pages). How this typo undercuts Meyer's credibility is anybody's guess, especially since PubMed reveals well over a dozen papers with the "splicesome" spelling in various science journals.[54] For that matter, if perfect copy-editing is to be the test of scientific rigor, Matheson himself should watch out: There are typos on his own blog (in one instance he forgot to insert an "o" into the name of Dr. Russell "Dolittle," much as Meyer forgot an "o" in "spliceosome").

---

53. See Rubins, J.B., et al., "Distinct roles for pneumolysin's cytotoxic and complement activities in the pathogenesis of pneumococcal pneumonia," *American Journal of Respiratory and Critical Care Medicine* 153:4 (1996), pp.m1339–1346 or Pneumolysin entry at: http://www.uniprot.org/uniprot/P0C2J9.

54. For some examples, see Hieda, M., et al., "Nuclear Import of the U1A Spliceosome Protein Is Mediated by Importin α/β and Ran in Living Mammalian Cells," *Journal of Biological Chemistry* 276 (2001), pp. 16824–16832; Lidie, Kristy B. and Frances M. Van Dolah, "Spliced Leader RNA-Mediated trans-Splicing in a Dinoflagellate, *Karenia brevis*," *Journal of Eukaryotic Microbiology* 54:5 (2007), pp. 427–435 (2007); Lester, Leo, Andrew Meade, and Mark Pagel, "The slow road to the eukaryotic genome," *BioEssays* 28:1 (2005), pp. 57–64.

MATHESON ALSO FAULTS Meyer for claiming that RNA splicing is accomplished using not just the spliceosome but also exonucleases and endonucleases. Matheson writes that Meyer is wrong because the spliceosome "is not known to include either exonucleases or endonucleases." But Matheson didn't read Meyer correctly. Meyer doesn't say the spliceosome contains exonucleases or endonucleases. He just says that, along with the spliceosome, they are involved in the process of correctly identifying and excising introns. And in fact, there's evidence that both endonucleases[55] and exonucleases[56] can be involved in the splicing process.

Since Matheson is so fastidious about scientific accuracy when it comes to the spliceosome, one should point out his own overstated claim that the spliceosome "is made mostly of RNA." According to a 2009 paper in *PNAS*, "[t]he spliceosome is a massive assembly of 5 RNAs and many proteins"[57]—another paper suggests "300 distinct proteins"![58] So it seems the spliceosome is certainly made of RNA, but is not necessarily "made mostly of RNA."

Arguably Matheson's most vitriolic attacks on Meyer's book come during his discussion of "junk DNA," a discussion that is out of date.

In *SITC* Meyer writes that "the original DNA text in eukaryotic organisms has long sections of text called 'introns' that do not (typically) encode proteins. Although these introns were once thought to be nonfunctional 'junk DNA,' they are now known to play many important functional roles in the cell." (p. 125) For making this argument, Matheson accuses Meyer of "some combination of ignorance, sloth, and duplicity" and "layers of dishonesty," alleging "This is the discredited creationist 'junk DNA' ploy."

Discredited? In 2003 *Scientific American* addressed a striking rebuke to those who claim introns are genetic junk: "The failure to recognize the

---

55. See Xue, Song, Kate Calvin, Hong Li, "RNA Recognition and Cleavage by a Splicing Endonuclease," *Science* 312:5775 (2006), pp. 906–910.
56. See Staley, Jonathan P. and John L. Woolford Jr., "Assembly of ribosomes and spliceosomes: complex ribonucleoprotein machines," *Current Opinion in Cell Biology* 21:1 (2009), pp. 109–118.
57. Butcher, Samuel E., "The spliceosome as ribozyme hypothesis takes a second step," *Proceedings of the U.S. National Academy of Sciences* 106:30 (2009), pp. 12211–12212.
58. Nilsen, Timothy W., "The spliceosome: the most complex macromolecular machine in the cell?," *BioEssays* 25 (2003), pp. 1147–1149.

importance of introns 'may well go down as one of the biggest mistakes in the history of molecular biology.'"[59]

Matheson's retort is that functions have been uncovered for only a "handful" of introns. But recent data shows evidence of mass functionality. An April 1, 2010, article in *Nature* reported that "Biology's new glimpse at a universe of non-coding DNA—what used to be called 'junk' DNA—has been fascinating and befuddling. Researchers from... ENCODE showed that in a selected portion of the genome containing just a few per cent of protein-coding sequence, between 74 percent and 93 percent of DNA was transcribed into RNA."[60] A variety of papers indicate that huge portions of DNA is being transcribed, hinting at function.[61] Indeed, introns can affect gene expression even when they're *not* transcribed.[62] If Matheson ever finishes reading *SITC*, he'll find a long list of functions discovered for noncoding DNA, citing over 45 papers from the mainstream scientific literature (p. 407). Even the journal *Science* stated that the "junk DNA" mindset has "repelled mainstream researchers from studying noncoding DNA,"[63] refuting Matheson's claims to the contrary elsewhere.[64]

---

59. Gibbs, Wayt T., "The Unseen Genome: Gems Among the Junk," *Scientific American* 289:5 (2003), p. 47.
60. Hayden, Erika Check, "Life Is Complicated," *Nature* 464 (2010), pp. 664–667.
61. See Mattick, John S. and Igor V. Makunin, "Non-coding RNA," *Human Molecular Genetics*, 15 (2006), pp. R17–R29; Ying, Shao-Yao, Donald C. Chang and Shi-Lung Lin, "MicroRNA (miRNA): Overview of the RNA Genes that Modulate Gene Function," *Molecular Biotechnology*, 38 (2008), pp. 257–268; Dinger, Marcel E., Paulo P. Amaral, Timothy R. Mercer, and John S. Mattick, "Pervasive transcription of the eukaryotic genome: functional indices and conceptual implications," *Briefings in Functional Genomics and Proteomics* 8 (2009), pp. 407–423; Tsirigos, Aristotelis and Isidore Rigoutsos, "Alu and B1 Repeats Have Been Selectively Retained in the Upstream and Intronic Regions of Genes of Specific Functional Classes," *PLoS Computational Biology* 5:12 (2009), e1000610; Louro, Rodrigo, Anna S. Smirnova and Sergio Verjovski-Almeida, "Long intronic noncoding RNA transcription: Expression noise or expression choice?," *Genomics* 93 (2009), pp. 291-298; Shomron, Noam and Carmit Levy, "MicroRNA-Biogenesis and Pre-mRNA Splicing Crosstalk," *Journal of Biomedicine and Biotechnology* 2009 (2009), 594678.
62. Swinburne, Ian A. and Pamela A. Silver, "Intron Delays and Transcriptional Timing During Development," *Developmental Cell* 14 (2008), pp. 324–330.
63. See Makalowski, Wojciech, "Not Junk After All," *Science* 300 (2003), p. 5623 (emphasis added).
64. See http://sfmatheson.blogspot.com/2008/01/talking-trash-about-junk-dna-lies-about.html. For a further rebuttal to Matheson, see also Cornelius Hunter's blog, "Stephen Matheson: Talking Trash About Junk DNA," at http://darwinsgod.blogspot.com/2009/10/stephen-matheson-taking-trash-about.html (October 3, 2009). Matheson responds to Hunter

21. Gotcha! Stephen Meyer's Spelling & Other Weighty Criticisms

Are all these scientists part of a nefarious plot to promote, as Matheson calls it, a "discredited creationist 'junk DNA' ploy"?

Arriving at Meyer's Chapter 8, Matheson actually concedes that it is "pretty good," but objects that "a designer can put chance occurrences to very good use," thus an event occurring due to chance does not negate design. He gives the example of a designer using a chance-based coin toss to determine some decision. However, rather than negating the distinction between chance and design, this example highlights it.

Consider a coin toss before a football game to determine who kicks off. Here, an intelligent referee will deliberately use an undirected chance-based event—but that's precisely because the referee desires an event where he has no control over the outcome. The referee's decision to flip a coin doesn't mean the referee is now directing the outcome of the coin toss. The outcome of the coin toss is still undirected, the result of chance. The distinction between chance and design still stands.

Matheson's discussion of Chapters 9 and 10 begins with another tedious parade of slurs: He claims the chapters "advance a straw man so idiotic that I wonder whether Meyer will be able to reclaim any significant intellectual integrity in the chapters that follow." He charges Meyer with having "purely propagandistic aims," which "do serious damage to the book's credibility and to the author's reputation."

What disreputable blunder did Meyer make this time? Turns out Meyer's crime is observing that some theorists attributed the origin of life to "chance," a hypothesis Matheson doubts was ever actually put forth. You read that right.

Of course Meyer cites multiple authorities from the origin of life research community—giants such as Francis Crick or George Wald (see p. 195)—advancing the hypothesis that life arose by "chance" or "accident." But this isn't enough to convince Matheson. I'm sure that theorist David Deamer's suggestion that "genetic information more or less came out of

---

merely by calling him "a poorly equipped ID demagogue." See http://sfmatheson.blogspot.com/2009/12/resurrection.html.

nowhere by chance assemblages of short polymers"[65] wouldn't convince Matheson either. How about the more charitable reader?

Had Meyer stopped *SITC* at Chapter 10 then perhaps Matheson could say Meyer advocates a "straw man." But these chapters are by no means Meyer's entire argument. Meyer's rhetorical structure is to first assess the "chance" hypothesis—but he fully acknowledges that there are more sophisticated theories to be dealt with later in the book which use various combinations of chance and law, including natural selection. Meyer thus writes at the close of Chapter 10:

> Some theorists, notably those proposing an initial "RNA world," have sought to retain a role for chance by suggesting that natural selection might have played a key role in the origin of life, even before the origin of a fully functioning cell. They propose combining chance with natural selection (or other lawlike processes) as a way of explaining how the first cell arose. In doing so, they argue that random processes would have had to produce much less biological information by chance alone. Once a self-replicating molecule or a small system of molecules had arisen, natural selection would "kick in" to help produce the additional necessary information. In Chapter 14, I evaluate theories that have adopted this strategy. (pp. 227-228)

Meyer by no means leaves his readers hanging with the impression that materialists must believe an entire living cell appeared all-at-once by "chance," directly negating Matheson's criticism.

Matheson opens his review of Chapters 9 and 10 by defining "straw man," claiming that this is what Meyer puts forth. Like a judge who issues a verdict after only reviewing half the evidence, Matheson is prematurely accusing Meyer of misrepresenting origin of life thinking. What's ironic is that by accusing Meyer of creating a straw man and ignoring *SITC*'s much more comprehensive argument, it's *Matheson* who is promoting the straw man.

Is this really the best critique possible from someone who is actually reading *SITC*?

---

65. David Deamer, quoted in Susan Mazur, *The Altenberg 16: An Expose of the Evolution Industry*, p. 180 (Scoop Media, 2009).

22. Matheson's Intron Fairy Tale

Richard Sternberg

> The failure to recognize the importance of introns "may well go down as one of the biggest mistakes in the history of molecular biology."
> —John Mattick, *Scientific American*, November 2003

On Friday, May 14, 2010, I watched as Steve Meyer faced his critics—two of them anyway, Art Hunt and Steve Matheson—at Biola University in Los Angeles.

Matheson had previously claimed that Meyer misrepresented introns in his book, *Signature in the Cell*. (Introns are non-protein-coding sequences of DNA that occur within protein-coding regions.) In a blog post dated February 14, Matheson had accused Meyer of "some combination of ignorance, sloth, and duplicity" for stating in his book that although introns do not encode proteins they nevertheless "play many important functional roles in the cell."[66]

Calling Meyer's statement "ludicrous," Matheson wrote on his blog that biologists have identified functional roles for only "a handful" of the 190,000 or so introns in the human genome:

> How many? Oh, probably a dozen, but let's be really generous. Let's say that a hundred introns in the human genome are known to have "important functional roles." Oh fine, let's make it a thousand. Well, guys, that leaves at least 189,000 introns without function.

Matheson added that "there are more layers of duplicity in the 'junk DNA' fairy tale than Meyer has included in his book," which (Matheson concluded) uses science to advance an agenda in which "rigorous scientific truth-telling is secondary."

Naturally, I expected Matheson to bring up this devastating criticism at the Biola event on May 14. But he said nothing about Meyer's "ludicrous"

---

66. http://sfmatheson.blogspot.com/2010/02/signature-in-cell-chapters-4-and-5_14.html

notions of intron functions that evening, and he was mum about all the other layers of duplicity that he claims to be privy to. This was probably wise, because Matheson is wrong about intron functionality.

The segments of our DNA that are commonly called "genes" consist of protein-coding exons and non-protein-coding introns. Initially, the entire DNA segment is transcribed into RNA, but between ninety and ninety-five percent of the initial RNAs are "alternatively spliced."

What is alternative splicing? Imagine that the initial RNA derived from its DNA template has the organization A—B—C—D—E—F, where the letters represent blocks that specify amino acid sequences and the dashes in between the letters stand for introns. Alternative splicing enables multiple proteins to be constructed given the same RNA precursor, say, ABCDF, ACDEF, BCDEF, and so forth. In this way, hundreds or thousands of proteins can be derived from a single gene.

There's more. The messenger RNAs that are produced by this process—and therefore the proteins that are made in a cell—are generated in a way that depends on the stage of development as well as the cell and tissue type. In the above example, a nerve cell may express the ACDEF version of a messenger RNA whereas a pancreatic cell may produce only the BCDE version. The differences are biologically essential.

What does this have to with introns? Everything. It is the *presence of introns* that makes this permutative expansion of messenger RNAs possible in the first place.

So let's do the math. At least ninety percent of gene transcripts undergo alternative splicing, and there are at least 190,000 introns in the human genome. That means we have at least 0.90 x 190,000 = 171,000 introns that participate in the alternative-splicing pathway(s) available to a cell.

Someone could argue that the sequences directing alternative splicing are in the protein-coding regions of the RNA. That is, one could argue that while introns do indeed make splicing possible, they are merely "junk" fillers, with the exons indicating where and how the spliceosome is to do its cutting and pasting. Yet such an argument would be false. In order for alternative splicing to work properly, it is necessary not only that exons be demarcated from introns, but also that the splicing process be correctly modulated.

And introns contribute significantly to such modulation. How do I know this? Take a look at figure 3a in a 2010 article in *Nature* about alternative-splicing regulation in mice.[67] The figure provides the tabulated distribution of alternative splicing code motifs within three generalized exons (the white, purple, and orange blocks on the top) and two introns (the thin broken, black/light blue and dark blue/olive brown lines). At the top of each column, four letters indicate the tissues from which the data were derived: C = central nervous system; M = muscle; E = embryo; and D = digestive system. It is clear that introns are just as rich in splicing-factor recognition sites as are exons. The authors of the article—titled "Deciphering the splicing code"—conclude that the evidence "predicts regulatory elements that are deeper into introns than previously appreciated."

Using the mouse as a surrogate, we can infer that the roughly 171,000 human introns involved in alternative splicing probably have a similar distribution of formatting codes, which are necessary to ensure that the proper proteins are made at the correct developmental stage and in the appropriate cells and tissues. Even if we were off by a factor of two, we would still be left with 85,500 introns that function in the process of alternative splicing.

This is the tabulated distribution of alternative splicing code motifs within three generalized exons (the white, purple, and orange blocks on the top) and two introns (the thin broken, black/light blue and dark blue/olive brown lines). This figure was taken from a 2010 article in *Nature* about alternative-splicing regulation in mice. At the top of each column, four letters indicate the tissues from which the data were derived: C = central nervous system; M = muscle; E = embryo; and D = digestive system. It is clear that introns are just as rich in splicing-factor recognition sites as are exons. The authors of the article—titled "Deciphering the splicing code"—conclude that the evidence "predicts regulatory elements that are deeper into introns than previously appreciated."

---

67. The figure is in color and so cannot be accurately reprinted here, but it can be viewed online at http://www.nature.com/nature/journal/v465/n7294/fig_tab/nature09000_ft.html. The article in which the figure appears is Barash, Y., et al., "Deciphering the splicing code," *Nature* 465:53–59 (2010).

Using the mouse as a surrogate, we can infer that the roughly 171,000 human introns involved in alternative splicing probably have a similar distribution of formatting codes, which are necessary to ensure that the proper proteins are made at the correct developmental stage and in the appropriate cells and tissues. Even if we were off by a factor of two, we would still be left with 85,500 introns that function in the process of alternative splicing.

This is not the only evidence that Matheson ignored. For example, non-translated microRNAs regulate the developmental expression of messenger RNAs, and small nucleolar RNAs are essential for the processing of ribosomal RNAs (which in turn are essential for protein production). The human genome contains 1,664 known genes for the former and 717 known genes for the latter, and *the majority of these genes occur in introns.*

Then there are the regulatory codes associated with such RNA genes, which also occur in introns. And RNAs that emanate from introns but that are not part of messenger RNAs; 78,147 of them are known to exist in humans. Even if only 10 percent of the latter RNAs play some role in cellular organization, we have far more than "a handful" of functional introns in this category alone.

And there's still more. RNA is essential for chromatin organization in the nucleus. When chromatin-associated RNA is degraded by experimental means, the geometry of chromosomes and nuclear metabolism is adversely affected. Yet a recent study of this class of RNAs in human cells revealed that over half of the transcripts (52 percent) are derived from... introns!

I could go on. Various DNA control modules have been mapped to introns, including alternative promoters, enhancers, silencers, and nuclear matrix attachment sites—some of which influence genes that are located over a million base pairs away on the chromosome. But sorting through all the studies that have been published on this subject would be a big job.

A job, obviously, that Matheson has not done—though whether through ignorance, sloth, or duplicity I cannot say.

23. Let's Do the Math Again

Richard Sternberg

In the preceding chapter, I criticized Calvin College biologist Steve Matheson's incorrect view of "junk" DNA. Matheson in his blog had argued in February that the human genome contains about 190,000 introns (stretches of non-protein-coding DNA that interrupt protein-coding genes), of which "only a handful" had important functional roles. "How many? Oh, probably a dozen," he wrote, "but let's be really generous. Let's say that a hundred introns in the human genome are known to have 'important functional roles.' Oh *fine*, let's make it a thousand."

On the contrary, I pointed out that at least 90 percent of genes are alternatively spliced, meaning that 0.9 x 190,000 = 171,000 introns are involved in alternative splicing, an essential process that helps to ensure that the proper proteins are made at the correct developmental stage and in the appropriate cells and tissues.

Along comes University of Toronto biochemist Larry Moran, an outspoken Darwinist who hates the Center for Science and Culture so much he would probably heap abuse on us for saying that the Earth goes around the Sun. Sure enough, Moran wasted no time jumping on me for being an "Intelligent Design Creationist." He posted the relevant portion of my critique and concluded: "It's up to you, dear readers, to figure out all the things wrong with this explanation. You can start with the math. Arithmetic isn't one of their strong points."[68]

So let's do the math. Again. I will make the task easy for everyone—even Moran and Matheson:

Step 1. There are ~25,000 protein-coding genes in the human genome.

Step 2. There are 190,000 introns/25,000 protein-coding genes = 7.6 introns/gene on average.

---

68. http://sandwalk.blogspot.com/2010/06/creationists-introns-and-fairly-tales.html

Step 3. Ninety percent (possibly more) of gene transcripts undergo alternative splicing. Hence, 0.9 x 25,000 = 22,500 genes (actually, their RNAs) undergo alternative splicing.

Therefore, 22,500 genes x 7.6 introns/gene = 171,000 introns involved in alternative splicing.

This is just a rough estimate, of course. And as I wrote in my original critique of Matheson, even if I'm off by a factor of two we are still left with far more functional introns than Matheson acknowledges. This compels me to ask Steve Matheson: How exactly did you come up with your estimates? And what about you, Larry Moran? What sort of arithmetic are you using?

24. Darwinian Tree-Huggers: You Gotta Love Their Devotion

Douglas Axe

STEVE MEYER RECENTLY GAVE A LECTURE SUMMARIZING THE ARGUments put forward in his book *Signature in the Cell* to an audience of 1,400 (including me) at Biola University. After Steve sat down, two of his critics, Steve Matheson and Arthur Hunt, were invited to put their questions to him.

Matheson and Hunt both referred to my work and to Meyer's use of it, Matheson having since posted his points on his blog. As is often the case when complex subjects are debated in front of an audience, things got a bit muddled. I stood up at one point with the intent of commenting but wasn't able to get the attention of the moderator, so I'll comment here instead.

The specific work in the *Journal of Molecular Biology* to which Meyer, Matheson and Hunt referred has added to the scientific case for functional protein sequences being extraordinarily rare within the whole space of possibilities. Matheson started off by arguing not that this deduction of extraordinary rarity is incorrect, but rather that it is irrelevant to the debate between Darwinism and Design. According to him, "What is relevant is whether the protein's place in sequence space is linked through achievable steps to other points in sequence space" in a manner traditionally represented by Darwin's branching tree. His reasoning seems simple enough:

> I pointed out that a standard evolutionary account of that tree, whether it's a tree of species or a tree of people or a tree of proteins, makes no prediction about the rarity (or commonness) of function or adaptation within the space that the tree inhabits. In the case of proteins, the branches of the tree are particular proteins, and the proteins are linked to each other by common ancestry. Whether each

branch represents a fantastically rare structure that has a function, or just represents one choice among zillions of [comparably functional] alternatives, is really not relevant to the question of how the protein's structure came to be.

But this is a naïve understanding of prediction. Theories have both consequences and assumptions. If a theory is correct, then its consequences will prove true when tested, and so will the assumptions on which it is predicated. The claim that a theory is correct therefore amounts to a prediction both that its consequences will prove true and that its assumptions will prove true.

Tree-like relationships are what Darwinian evolution produces—they are the consequence of its operation. But if Darwinism is to work as a theory of origins, it must explain not just trees but trees with remarkable transformations of form and function scattered throughout their branches. For a century and a half *this* has been the major point in dispute—whether such remarkable transformations can possibly happen through small adaptive steps. If as a rule they can, then Darwinism works. If as a rule they can't, then Darwinism flops.

Matheson recognizes the immediate assumption on which the Darwinian account of proteins rests—that the whole set of biological proteins must be "linked through achievable steps"—but he doesn't seem to see what would have to be true in order for that to be true. Assumptions rarely travel alone.

P ART OF THE difficulty is that the degree of rarity we're talking about here is so far beyond our everyday experience that our intuitions tend to be unreliable. When we think of extraordinarily rare events, we think of winning the lottery or being struck by lightening, both of which are actually very common events on the scale relevant to protein origins.

Picture this instead. Suppose a secretive organization has a large network of computers, each secured with a unique 39-character password composed from the full 94-charater set of ASCII printable characters. Unless serious mistakes have been made, these passwords would be much uglier than any you or I normally use (and much more secure as a result). Try memorizing this:

C0$lhJ#9Vu]Clejnv%nr&^n2]B!+9Z:n`JhY:21

Now, if someone were to tell you that these computers can be hacked by the thousands through a trial-and-error process of guessing passwords, you ought to doubt that claim instinctively. But you would need to do some math to become fully confident in your skepticism. Most importantly, you would want to know how many trials a successful hack is expected to require, on average. Regardless of how the trials are performed, the answer ends up being at least half of the total number of password possibilities, which is the staggering figure of 10 raised to the power 77 (written out as 100,000,000,000,000,000,000,000,000,000,000,000,000,000,000,000,000,000,000,000,000,000,000,000,000,000,000,000). Armed with this calculation, you should be very confident in your skepticism, because a 1 in 10^{77} chance of success is, for all practical purposes, no chance of success.

My experimentally based estimate of the rarity of functional proteins produced that same figure, making these likewise apparently beyond the reach of chance. So, with due caution, let's transfer Matheson's reasoning from the problem of protein origins to this hypothetical hacking problem. It is as though Steve Meyer has said that computers with passwords of the strength described above cannot be hacked by trial and error, and Steve Matheson has responded that password strength has nothing to do with it.

That's a peculiar response. Reading between the lines, I suspect the train of thought is something like this: We know that there are millions of computers on the organization's network, not just the thousands that are to be hacked, and we know that the network is arranged in a branching pattern with neighboring machines having passwords that differ by only one character, so hacking the first machine will make it easy to hack the rest.

Ummm... but we *don't* really know these things. I can understand why Darwinists *presume* the equivalent things to be true for proteins (and even *want* them to be true), but Darwinism is itself the thing in question here, so all its presumptions need to be set aside.

Certainly an IT manager *could* configure a network in such a highly hacker-friendly way, if that were the objective. But absent any reason to think this was the objective, it would be a mistake to presume so. All we really know is that there are thousands of machines to be hacked and that they all use 39-character passwords. The only sensible deduction under these circumstances is that every attempt to hack one of these machines by sampling passwords must fail.

It seems to me that the default assumption for proteins ought to follow the same generalization—that fantastically rare points in vast spaces don't line up like stepping stones unless something forces them to. Might there be such a force for proteins—even a non-teleological one? Conceivably. So, it would be perfectly reasonable to ask whether something might possibly force functional protein sequences to align in this way. But to dismiss their fantastic rarity as irrelevant, as Matheson has done, is to misunderstand the problem entirely.

In fact, although Matheson is right that my prior paper focused on the rarity of functional protein sequences rather than on their isolation, a recent paper in *BIO-Complexity* examines directly the implications of their extreme rarity for protein evolution. Rarity is by no means the only aspect of the problem that has to be considered, but it certainly is a key aspect, and in the final analysis it appears to be the decisive one.

25. Is Intelligent Design Bad Theology?

Jay Richards

OVER THE YEARS, PROPONENTS OF INTELLIGENT DESIGN HAVE spent much of their time developing the theoretical tools for inferring design and developing the empirical case for design in fields such as cosmology, astronomy, origin of life studies, and molecular biology. In contrast, many critics have spent their time attacking the supposed theology behind ID.

In the last few weeks, *The Guardian* (in the UK) has been publishing responses to the following question: "Is Intelligent Design Bad Theology?" Philosophers Michael Ruse and Stephen Fuller have weighed in on the question. Recently, journalist Mark Vernon responded to the question[69] by "reviewing" Stephen Meyer's book, *Signature in the Cell*. Based on his interpretation of Meyer's argument, Vernon concludes that ID is "bad science, bad theology, and blasphemy." That puts it strongly. Unfortunately, Vernon's strong language is not supported by strong arguments.

Surprisingly, Vernon's brief summary of Meyer's argument is actually pretty good; but then he quickly goes off the rails. His complaint, initially, is that Meyer's argument leads to the conclusion that ID is the best explanation for the origin of life to date; but, "in truth, no one really knows what life is, let alone how it arose. The work in the last half century or so on DNA has only deepened the problem—vastly deepened it."

The obvious response is, So what? As Meyer argues in his book, there is far more to life than the little bit we know at the moment. Meyer argues that there is far more information in a cell, for instance, than is present in the coding regions of DNA. But Meyer's argument is based squarely on what we do know about life and its informational properties, not on what we don't

---

69. http://bio-complexity.org/ojs/index.php/main/article/view/25

know. Vernon seems to think that if we don't know everything about life, any argument based on what we do know will be an argument from ignorance. This is bizarre. Such curious "reasoning," if applied consistently, would mean we could never make arguments or draw conclusions about anything, since there would always be something we don't know. The only thing we could do is remain silent. Frankly, I don't think Vernon means what he says here. If he did, he would be giving the same advice to everyone, and not just to ID proponents.

As it is, everyone is in the same boat. Good arguments will be based on what we know at the moment. And that's exactly what Steve Meyer does in *Signature in the Cell*.

Building on the argument described above, Vernon then proceeds to his theological complaint. Although his critique is directed officially to *Signature in the Cell*, it becomes clear that he intends the critique to apply to ID more generally. An early sign that his critique will misfire is his reference to "Newton's view of the universe" as "a deistic belief in a divine architect." Only problem: Newton was not a deist. Deism is the view that God starts the world on its course and then doesn't interact any more with it.

Newton, in contrast, thought God not only set up the world at the beginning, but also constantly upheld and interacted with it in a variety of ways. He was a harsh critic of Cartesians who seemed to consign God to a place only at the cosmic beginning. Whatever one makes of Newton's specific views, they were a far cry from deism.

Vernon faults ID for similar inaccuracy in "assuming that God could be a scientific explanation at all. To do so has long been observed to be ridiculous." Unfortunately, he doesn't cite any sources from the ID literature to substantiate his characterization of ID. That's hardly surprising, since ID proponents have explained over and over and over again that ID per se isn't committed to a specific mode of divine causality. ID is about detecting the effects of intelligent agency within nature (divine or otherwise). Either there is evidence for such effects within nature, or there is not. Detecting the effects of design is different from specifying how the design is implemented, or by whom.

25. Is Intelligent Design Bad Theology?

Apparently oblivious to these distinctions, Vernon tries to seal ID off in a "religious" compartment. "Belief and science are two different kinds of explanation, one moral, the other material," he explains. "Explanations based on 'belief' have to do with morals, not science." To insert one type of explanation in place of the other, according to Vernon, is to make a category mistake.

Now let's set aside the fact that he's confusing an argument for agency in explaining something in nature with religious belief, and just focus on what he says about the nature of religious belief. It's clearly false. Even the most superficial student of religion knows that various religions, such as Judaism, Hinduism, Christianity and Islam, intend to explain all sorts of things about the world. No religion is obligated to restrict its explanations to morality, and few have done so. So as a description of what real religions actually do, Vernon's assertion is baseless.

Based on his analysis of "scientific" and "religious" explanations, Vernon concludes that ID is bad theology. Indeed, he claims that it's blasphemy, because it purportedly invokes God to explain something in the world:

> God is something else again, which Thomas Aquinas, the medieval theologian, explored in the notion that creation is "out of nothing." The "*ex nihilo*" is not supposed to be a demonstration of God as a scientific whiz-kid, so amazing that he doesn't even need matter to make the cosmos. Rather, it's to say that the universe was created with no instrumental cause. It is the original free lunch, offered purely out of God's love. You can argue about whether you'd have picked what's on the menu. But to insert God into the causal chain is a category mistake and, in fact, technically a blasphemy. It implies that God is one more thing along with all the other things in the universe. You're not dealing with divinity there, but an idol.

So ID proponents are guilty of both blasphemy and idolatry.

What to say? Well, it's clear that Vernon is confusing a cartoonish stereotype of ID with the real thing. No ID theorist has ever said: "Insert God here." ID theorists offer detailed arguments for why intelligent agency is the best explanation for various features of the natural world—or of the natural world itself. For instance, Steve Meyer goes into extraordinary detail in *Signature in the Cell* explaining why chance, mere self-organization, or chance plus a blind selection mechanism are inadequate to explain biological information. He also provides detailed positive arguments for why we should attribute such information to intelligent design. His argument has clear theological implications, but it doesn't rest on narrow theological premises. He simply asks that intelligent design be considered a possible explanation.

But let's set aside the details about ID and consider Vernon's theological assertion on its own terms.

Let's imagine someone who does explicitly invoke God in explaining some feature of nature, someone like Thomas Aquinas. Does "inserting God into the [natural] causal chain" commit "a category mistake" and make one guilty of "blasphemy"? Would it imply that "God is one more thing along with all the other things in the universe"? Specifically, would such a claim contradict a fundamental principle of Christian theology? No, of course it wouldn't.

Christianity has traditionally taught that God is omnipotent, free and sovereign over his creation. God is qualitatively more powerful than mere human beings. He can do far more than human beings, not less. Since human beings, despite our limitations, can build 747s, there's nothing preventing God from doing the same (though we have no reason to think he has done so).

Vernon is trying to use the doctrine of creation *ex nihilo* as a catch-all, to suggest that the doctrine somehow bars God from acting in other ways within the universe. There's no basis whatsoever for this move in Christian theology. It's invented from whole cloth. The fact that God created the universe *ex nihilo* doesn't mean that that's his only way of acting. The only justification I can think of for limiting God's freedom to act within the created order would be to square Christian theology with naturalism. But then it would cease to be Christian theology.

In reality, Christianity is firmly committed to God doing all sorts of things within the created order. According to Christian theology (which is relevant since Vernon appeals to Thomas Aquinas), God creates the world from nothing, he raises people from the dead, he became incarnate as a human being, he caused Mary to become pregnant without the benefit of a human male, and so forth. If the latter claim is true, then the proper explanation for Mary becoming pregnant is the direct causality of God within the natural order.

Every educated Westerner, whether believer or unbeliever, knows perfectly well that Christians believe that God is both the creator of everything that is, and that he acts within nature. In fact, it's hard to think of a less controversial claim. Richard Dawkins and the Archbishop of Canterbury both know this. So it's just silly for Vernon to assert that invoking God as a cause within nature is "blasphemous."

WHAT ABOUT HIS assertion that invoking divine causality within nature somehow makes God "one more thing along with other things in the universe"?

Unfortunately, this is just an assertion. Vernon doesn't provide even a pretense of an argument. And it's hard to think of any argument in its favor. If God is free and sovereign over his creation, then he can do what he wants to do. He's under no obligation to conform to Mark Vernon's rules of tidiness and propriety. If he wants to act directly within the created order for his own purposes, he can certainly do that. And in so doing, God doesn't become "one more thing along with all the other things in the universe." He continues to be God. Vernon is confusing cause with effect. God may act directly in the created order, and the effect of his action would become part of that order. But that doesn't mean God therefore becomes merely one more member of the universe.

Of course, the claim that God acts directly in the created order might seem blasphemous to a theology that has fully capitulated to naturalism, such as the deism that Vernon falsely attributes to Newton. Vernon is free to defend such a theology, and to define everyone who claims that God can act within nature as a blasphemer. But in that case, he should explain that according to his view of God, every traditional theist on the planet is guilty of blasphemy. And he should distinguish such anti-deistic "blasphemy" from ID, which doesn't entail a specific mode of divine causality.

About the Authors

Douglas Axe

David Axe is the director of Biologic Institute. His research uses both experiments and computer simulations to examine the functional and structural constraints on the evolution of proteins and protein systems. After a Caltech PhD he held postdoctoral and research scientist positions at the University of Cambridge, the Cambridge Medical Research Council Centre, and the Babraham Institute in Cambridge. His work has been reviewed in *Nature* and featured in books, magazines and newspaper articles, including *Life's Solution* by Simon Conway Morris, *The Edge of Evolution* by Michael Behe, and *Signature in the Cell* by Stephen Meyer.

David Berlinski

David Berlinski is a senior fellow in the Discovery Institute's Center for Science & Culture. He is the author of numerous books, including *The Devil's Delusion: Atheism and Its Scientific Pretensions* (Crown Forum), *Infinite Ascent: A Short History of Mathematics* (Modern Library), *The Secrets of the Vaulted Sky* (Harcourt), *The Advent of the Algorithm* (Harcourt Brace), *Newton's Gift* (Free Press), and *A Tour of the Calculus* (Pantheon). Berlinski received his PhD in philosophy from Princeton University and was later a postdoctoral fellow in mathematics and molecular biology at Columbia University. He lives in Paris.

David Klinghoffer

David Klinghoffer is a senior fellow at the Discovery Institute and the author of *The Lord Will Gather Me In: My Journey to Orthodox Judaism* (Free Press), *Why the Jews Rejected Jesus: The Turning Point in Western History* (Doubleday), *The Discovery of God: Abraham and the Birth of Monotheism* (Doubleday), and other books. He writes often for the *Los Angeles Times*, *National Review*, *The Weekly Standard*, the *Forward*, and other publications. He is a former literary editor of *National Review*.

Casey Luskin

Casey Luskin is an attorney with graduate degrees in both science and law. He earned his BS and MS in Earth Sciences from the University of California, San Diego. In his role at Discovery Institute, Luskin works as Program Officer in Public Policy and Legal Affairs, helping educators and policymakers nationwide to teaching evolution accurately. Luskin has published in both law and science journals, including *Journal of Church and State*; *Montana Law Review*; *Geochemistry, Geophysics, and Geosystems*; *Hamline Law Review*; *Liberty University Law Review*; *University of St. Thomas Journal of Law and Public Policy*; and *Progress in Complexity, Information, and Design*.

Stephen C. Meyer

Stephen C. Meyer is director of the Discovery Institute's Center for Science and Culture and a founder both of the intelligent design movement and of the Center for Science & Culture, intelligent design's primary intellectual and scientific headquarters. Dr. Meyer is a Cambridge University-trained philosopher of science, the author of peer-reviewed publications in technical, scientific, philosophical and other books and journals. His signal contribution to ID theory is given most fully in *Signature in the Cell: DNA and the Evidence for Intelligent Design*, published by HarperOne in June 2009. For more information on the book, and more about Dr. Meyer's views on intelligent design, visit his website at www.signatureinthecell.com.

Paul A. Nelson

Paul A. Nelson received his PhD from the University of Chicago in philosophy. He is a philosopher of biology, specializing in evo-devo and developmental biology. He has published articles in such journals as *Biology & Philosophy*, *Zygon*, *Rhetoric and Public Affairs*, and *Touchstone*, and chapters in the anthologies *Mere Creation*, *Signs of Intelligence*, and *Intelligent Design Creationism and Its Critics*.

Jay Richards

Jay Richards is a senior fellow of the Discovery Institute and a contributing editor of *The American* at the American Enterprise Institute. In recent years he has been a visiting fellow at the Heritage Foundation, and a research fellow and director of Acton Media at the Acton Institute. His most recent book is *Money, Greed, and God: Why Capitalism Is the Solution and Not the Problem* (HarperOne). Richards is also executive producer of several documentaries, including *The Call of the Entrepreneur*, *The Birth of Freedom*, and *Effective Stewardship* (Acton Media and Zondervan).

Richard Sternberg

Richard Sternberg is a research collaborator at the Smithsonian Institution's National Museum of Natural History. He joined Biologic Institute as a principal investigator in 2007. With expertise in evolutionary biology and bioinformatics, he studies the organization of genomic information and how it relates to organismal form. Holding PhDs in molecular evolution and in systems science, he has been a staff scientist at the National Center for Biotechnology Information and a research associate at the National Museum of Natural History, where he served as editor of the *Proceedings of the Biological Society of Washington*.

LaVergne, TN USA
15 November 2010
204987LV00013B/166/P